Anoectochilus yungianus

45th Anniversary Commemorative Edition

The Genera of Orchidaceae in Hong Kong

Shiu-ying Hu

The Chinese University of Hong Kong Press

The Genera of Orchidaceae in Hong Kong (45th Anniversary Commemorative Edition)
By Shiu-ying Hu

© The Chinese University of Hong Kong, 1977, 2022

ISBN: 978-988-237-269-6

Published by The Chinese University of Hong Kong Press
        The Chinese University of Hong Kong
        Sha Tin, N.T., Hong Kong
        Fax: +852 2603 7355
        Email: cup@cuhk.edu.hk
        Website: cup.cuhk.edu.hk

Printed in Hong Kong

Contents

FOREWORD TO THE COMMEMORATIVE EDITION by David T. W. Lau ix

FOREWORD by Gordon W. Dillon xiii

PREFACE by L. B. Thrower xv

Introduction xvii

I. THE BASIC FEATURES OF THE ORCHIDS IN HONG KONG 1
    Habit and Habitat 1
        1. Terrestrial Orchids; 2. Epiphytes; 3. Symbiosis
    Root 4
    Stem 5
    Leaves 6
    Inflorescence 6
    Flowers 7
    Fruit 16
    Seed and Seedling 17

II. KEYS AND DESCRIPTIONS 19
    Keys to the Subfamilies, Tribes, and Genera 19
    Descriptions of Genera with Illustrations and Keys to the Species 26

III. COMPOSITION AND PHYTOGEOGRAPHIC SIGNIFICANCE 125
    Systemic Summary 125
    Analysis of the Composition 129
        1. Generic Affinity with the Orchids of China; 2. Morphological Diversity;
        3. Small Species and Poor Populations; 4. Alphabetic List of the Species
    Phytogeographic Significance 136
        1. Indication of a Rich Flora and a Favorable Area; 2. Reference for Floristic
        Relationships; 3. Endemism at the Species Level
    Conclusion 138

IV. ORIGIN AND MEANING OF THE GENERIC NAMES OF HONG KONG ORCHIDS 141

GLOSSARY 149

BIBLIOGRAPHY 155

INDEX 159

MAJOR LIFE EVENTS OF PROFESSOR SHIU-YING HU 161

Foreword to the
Commemorative Edition

INTEGRATING TAXONOMY AND CUTTING-EDGE RESEARCH

When I came across the original manuscript of *The Genera of Orchidaceae in Hong Kong*, I realized that publishing a book 45 years ago was a lot more complicated than it is now. Every single word had to be typed out on a typewriter, and numerous drawings were needed in order to cover every part of each plant species. The description paragraphs and plant drawings had to be pasted onto paper manually in little pieces using small paper clips. Small amendments would change the layout of a whole page or even a few pages. These complicated procedures are hard to imagine nowadays, and it truly reflects the enthusiasm and dedicated efforts of the author, Prof. Shiu-Ying Hu.

The Genera of Orchidaceae in Hong Kong, written by Prof. Hu, was published in 1977 and was the first deliverable of the Flora of Hong Kong Project by The Chinese University of Hong Kong. The book is diverse in its content, with orchid habitats, growing habits, organ descriptions and authentication keys to sub-families, tribes, genera and species, as well as detailed descriptions of genus and species. In addition, sophisticated line drawings of plant organs and their dissected parts greatly enhance readers' understanding of the basic methodology and botanical evidence for taxonomic analysis. Prof. Hu made good use of plant structures, terminology and scientific drawings to illustrate the genuine identity of these orchid species. Hence, the content and approach of this book provides readers with a great deal of taxonomic knowledge. This type of knowledge transfer and educational approach was quite novel back in the 1970's and is greatly beneficial for both academics and taxonomists.

KNOWLEDGE TRANSFER

Prof. Hu's primary aim in publishing this book was to authenticate and document local plants, but she also made a detailed study of their genera distribution in mainland China and Southeast Asian countries. According to her analysis, 94% of the orchid genera in this book can also be found in Taiwan, the Philippines, Indonesia, and Malaysia, and other Southeast Asian regions. Another 30% of the genera have their ranges extending northward to the warm temperate region of the Yang-tze River, and 8% occurring even farther north to the temperate regions of China. Twenty-two percent were introduced from tropical America and Africa. This book therefore not only provides information on local orchids, but also is very valuable for studying orchids in nearby countries and regions. Prof. Hu's work highlights for me the importance of how local research and species documentation can also be a very good demonstration of how we can work with regional and larger areas for survey, documentation and collaborative research.

When determining the readership of this book, Professor Hu prepared it for orchid amateurs and taxonomists, and promoted it through orchid societies, international alliances, universities and secondary schools. She always put academic needs and public education first when considering the book's approach and content, in the hope that it would support the development of science, while at the same time promoting popular science. Such "knowledge transfer" and social applications of science are two essential ideas highly regarded by today's scientific researchers.

THE VALUE OF PLANT TAXONOMY NOWADAYS

Plant taxonomy is a kind of fundamental science subject which was recognized as core training in the last 200 years. New technology and tools in recent decades have been adopted for plant authentication and improving accuracy. DNA barcoding has been well adopted along with the discovery and usage of many barcoding regions, such as rbcL, ITS, matK, trnH-psbA. Nevertheless, it is noteworthy to understand that we still need to systematically collect raw plant samples to conduct voucher-based authentication and herbarium documentation. Therefore, flowers and fruit of the specimens are pre-requisites as evidence for plant authentication, which could definitely support the genuine results from DNA barcoding and genomic study; traditional morphological authentication remains irreplaceable by technology advancements. Hence, we have to preserve this traditional and basic science that helps nourish our curiosity to understand and appreciate nature. Furthermore, only by learning more about biodiversity and documenting it systematically can we make progress into more in-depth research.

Every pioneering work is usually full of difficulties and adventures; no one is really sure about the rewards or losses. Professor Hu's taxonomic work of plant collection and authentication, which she steadfastly carried out for more than 60 years, plus a lifetime of publications, truly reflected her work spirit, loyalty and capability. Her philosophy and life practices have motivated many young scientists and scholars to have more willpower and conviction to explore plant science and research. I strongly believe that this kind of traditional experience and knowledge—especially taxonomic knowledge—should be respected and held as a core value by scientific researchers, while at the same time making good use of new technologies and methods. In this way, scientific research, particularly that on plant authentication and its applications, will be even more innovative, fruitful and groundbreaking.

This commemorative edition, honoring the memory of Professor Hu, who died in 2012, is a reproduction of the 1977 edition, preserving the original format. "Major life events of Professor Shiu-ying Hu" has been included in the Appendix, highlighting some of the memorable moments that were important in Prof. Hu's life and to be remembered by us. This year is also the 10th anniversary of the Shiu-Ying Hu Herbarium. For the past ten years, we have been continuously studying and exploring plant taxonomy, and we will still be following in the remarkable footsteps of Prof. Hu in the days to come.

David T. W. LAU
Curator, Shiu-Ying Hu Herbarium, CUHK
Spring 2022

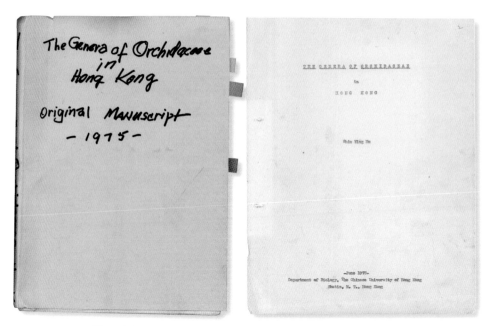

The original manuscript of *The Genera of Orchidaceae in Hong Kong.*

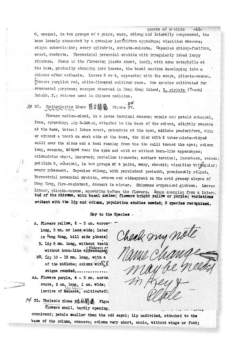

Dr. Hu's proofreading marks. Before word processors become common, every single change would mean a lot of meticulous hard work.

This newspaper clipping from 1975 is an interview with Professor Hu, in which the forthcoming publication of *The Genera of Orchidaceae in Hong Kong* was mentioned. Prof. Hu said: "To love and cultivate orchids is a kind of art, but to understand them, that is science."

Foreword

Southeast Asia is a vast geographical area whose northern and western boundaries may be roughly defined by a line extending from the Bay of Bengal north-ward in a great sweeping arc through the Gobi Desert and terminating at the northern tip of the Yellow Sea. Within this huge region there is great variety in natural resources, in topographical and climatic features, and in the richness of the flora and fauna, as well. It is a region noted for the diversity of its peoples and for its many ancient cultures; it is a region whose history during the past two hundred years has been shaken by social conflict, much of which derives from the impact of Western dynamism eroding the placid base of Eastern contemplative philosophies. It is a region whose peoples are now emerging into a new world which holds promise of blending the better elements of modern science and technology with the finer social and spiritual values of their traditional societies.

In the richly varied flora of this region, orchids are an important and sometimes conspicuous element, bolding interest for botanists and horticulturists alike. It is a fact that in China, and no doubt elsewhere in this region, orchids have a cultural history far antedating that of Europe and the Western world. In China, orchids were cultivated for centuries and prized not alone for their beauty and interest but for their presumed moral and inspirational values as well. Despite two thousand years of philosophical observation, however, there has been but little effort made to study the orchid flora in a modern botanical sense. Except for some of the work of Schlechter, Masamune, Tang and Wang, and a few others, no major taxonomic work on the orchids of China had been done until Dr. Shiu-ying Hu undertook her investigations which resulted in "The Orchidaceae of China," a series published in the Quarterly Journal of the Taiwan Museum, beginning 1971.

While politically separate, Hong Kong, comprising many offshore islands and a portion of the mainland, is floristically a portion of China. Thus this present work can be seen not only as a definitive study of the orchids of Hong Kong but also as a beginning to the more intensive study of the orchids of China itself.

Above this is the value this study offers in bridging a number of gulfs that separate people from one another: it links Hong Kong and China, it ties the scientific present to the cultural past, it provides a bridge between the botanist's study of a regional flora and the amateur's horticultural pursuit while both are specializing in this same fascinating family. Finally, and without doubt most importantly of all, it offers a positive bond between the East and the West, thus fulfilling in part that goal of orchidists everywhere—the eventual bringing about of "one world through orchids."

GORDON W. DILLON
Executive Secretary, American Orchid Society, Inc.
Botanical Museum of Harvard University
Cambridge, Massachusetts 02138, U.S.A.

Preface

The orchids are one of the most fascinating and attractive families of flowering plants. Their number of species, variety of growth habit, and the peculiarities of their flowers combine to make them so. These features also combine to make them one of the most difficult families for the systematic botanist.

Some sixty-four genera of orchids have been recorded for Hong Kong. So far as I am aware, no comprehensive account of them has been attempted previously. Consequently the present work by Dr. Shiu-ying Hu is a pioneering study.

This is a thorough-going scientific account of the genera of the orchids of Hong Kong, and is thus the base-line from which future taxonomists will begin their studies. But it is more than that, because Dr. Hu has taken considerable pains to make it serve the needs of the amateur, either experienced or beginner, who wishes to study local orchids. This has been done by augmenting the scientific descriptions with drawings of the growth habits of the plants, illustrating the important features of each genus with admirable line drawings, and providing a glossary to explain the technical terms. Moreover, a cross reference has been provided to the section on the Orchidaceae in the Check List of Hong Kong Plants compiled by the Hong Kong Herbarium (Agriculture and Fisheries Department).

I am sure that this work will provide information and encouragement for those who are already interested in local orchids, and will lead others to study them. Having said this I would emphasize that much can be learned without causing damage to individual plants or endangering a species. It is better to make a careful examination of the plants *in situ*, with photographs or a habit sketch and notes, than to pull it up by the roots; and two or three carefully selected flowers are sufficient for dissection and identification. In this way, all of us can help to ensure that Dr. Hu's work has the lasting value that it merits.

L. B. THROWER
Professor of Biology
The Chinese University of Hong Kong

Introduction

The purpose of publishing an illustrated treatise on the genera of Orchidaceae in Hong Kong is threefold: (1) to report a portion of the work accomplished by the Flora of Hong Kong Project; (2) to fill the vacuum of complete lack of references for a fundamental knowledge about this fascinating family for the people in China and the adjacent regions; and (3) to contribute our first hand observation of living specimens of orchids growing in Hong Kong for the information of orchidologists who depend on herbarium material for comparative studies elsewhere.

Flora of Hong Kong Project: The history, program, and progress of the Flora of Hong Kong Project was published in 1972.[1] The needs, cooperation and support of the orchidophiles of Hong Kong were important factors which made it possible for this publication to become one[2] of the first fruits of the Project. Needs were engendered by the fact that amateurs in Hong Kong depend upon the assistance of a professional taxonomist who works on the flora of the area to identify their specimens, explain the morphological characteristics of the species, and even to appreciate their achievements. A friendly relationship was established with the amateurs of orchids during my sojourn in Hong Kong. Among them are members of the Hong Kong Native Orchid Group under the leadership of Mrs. Gloria Barretto, and Dr. Sandy T. C. Lee, President of the Hong Kong Orchid Society. Members of the first group are interested in the exploration of the wilderness of the area, transplanting the spontaneous species of orchids, photographing the flowers, identifying them, and promoting their conservation. Members of the Orchid Society are interested in the importation, cultivation, hybridization, and selection of cultivars of horticultural merits. They have supplied me with flowering specimens for observation and illustration. Some rare species have been observed in the private collections of Dr. Tso Hsüeh-yên, and in the orchid garden of Kadoorie Farm.

A Supply of Reference: The appreciation and the cultivation of orchids have an ancient history, and a significant role in Chinese culture.[3] These had become

---

[1] Hu, S. Y. (1972) Floristic studies in Hong Kong. Chung Chi Journ. **11**:1–25. Maps 1–2, figures 1–3.

[2] Another manuscript, "Common and Useful Plants of Hong Kong," has been accepted for publication by The Chinese University of Hong Kong.

[3] Hu, S. Y. 1971. Orchids in the life and culture of the Chinese people. Chung Chi Journ. **10**:1–26. Map 1, figures 1–2.

popular in the Sung dynasty, and reached a highly developed stage in that period and the subsequent generations.

However, orchidology as a science is relatively new in China. Pioneers in this field of study are still among us. Orchidology, like many other disciplines of botanical science, had its beginning in the West. Amateurs as well as professional botanists in the East do not have the early literature published in Europe and America. Some of these publications concern the genera and species growing in the East. This lack of fundamental reference for a correct understanding and identification of their orchids forms a deep gorge which keeps the orchidophiles of the East from crossing to modern orchidology. Consequently their interest is limited to the cultivation and the appreciation of the cultivars. The few living collections of native orchids are not properly identified.

The present work is prepared to fill this gap in the lack of essential references on orchids, and to meet the need for a guide to identify the species growing in Hong Kong. Based on observations made with living specimens, in chapter I explanations are given to the basic features of orchids with 75 diagrams arranged in Figures 1–4 and with examples of local species. These represent the fundamental structures of 64 genera in 28 subtribes, four tribes and two subfamilies of Orchidaceae.

In chapter II keys and descriptions are prepared with terminology used by orchidologists throughout the world. For helping beginners to understand these terms, a glossary with Chinese equivalents and examples of included genera is provided. Visual aids for understanding the structure of each genus are given in form of habit sketches and analytical drawings of organs observed with the aid of dissecting microscope under strong illumination. All of the illustrations, with the exception of Figure 57, are original work. All the species known to me are listed in this work, and keys to taxa in genera which have more than one species are available. Four color plates (Figures 71–74) show the flowers of 18 species.

The usefulness of this work is not limited to the people of Hong Kong, for 94 percent of the genera included are represented in the spontaneous flora of Taiwan, the Philippines, and thence southward to Indonesia, Malaysia, and the mainland of southeastern Asia. Another 30 percent of the genera have their ranges extending northward to the warm temperate region of the Yangtze River, and 8 percent occurring even farther north to the temperate regions of China. Many of these genera also occur in Japan. Twenty-two percent of the genera are introduced from tropical America, Asia, or Africa. They are known in Hong Kong in cultivation only. For this reason, the material in this book provides a fundamental knowledge about orchids to people interested in the cultivated forms as well as to those enthusiatic with native species.

Amateurs and professional botanists are not satisfied in merely knowing the names of orchids. They would like to know why and how the species are in the vegetation of the area. Such material is presented in chapter III, on the composition and phytogeographic significance of the Orchidaceae in Hong Kong.

Advancement of Knowledge of Eastern Asian Orchids: Most genera and species of eastern Asian orchids were established on the basis of dry herbarium specimens deposited in one or few herbaria. It is well known that a dry orchid flower pressed flat appears very different from a fresh one. This condition has prevented the botanists working with herbarium specimens from seeing and reporting the delicate floral parts of the species they described. Consequently the information about these genera or species is limited to the insufficient material available to the orchidologists. The present work prepared from fresh flowering specimens gives a fully illustrated account of the delicate floral structure. In veery case except *Odontoglossum* and *Ornithochilus* where no flowering material was available, the description was made with the aid of a microscope before the specimen was turned over to the Botanical Artist, Mrs. Teresa Fung Wong, who has worked for eight years under my supervision. Both the description and illustrations were checked again with fresh material before the manuscript was submitted for publication. In the last three decades there has been a great deal of nomenclatural changes in Orchidaceae. Specialists differ in their concept of generic limits and specific interpretations. From material presented in this work, monographers may find the information on the exact structure of the genera and species growing in Hong Kong very helpful in their comparative study of all the species of a genus, a subtribe or a tribe.

ACKNOWLEDGEMENTS

I should like to take this opportunity to express my deep appreciation for help from all the various persons mentioned already in the introduction. I am indebted specially to Mrs. B. Walden for the paintings of *Anoectochilus yungianus* and *Cymbidium maclehoseae*, to Mr. James Leung for the color photographs (Figures 71–74), to Mr. Gordon W. Dillon, Executive Secretary of American Orchid Society for the Foreword, to Professor L. B. Thrower, Department of Biology, The Chinese University of Hong Kong for the Preface, to Dr. Stella L. Thrower, Department of Biology, The Chinese University of Hong Kong, Dr. Lily M. Perry, the Arnold Arboretum, Harvard University, and Dr. Herman R. Sweet, Orchid Herbarium of Oakes Ames, Botanical Museum, Harvard University, for going over the manuscript with many helpful suggestions, and to Mr. S. P. Lau, Keeper, and Mrs. T. P. Siu-Wong, Assistant, Hong Kong Herbarium, for the use of specimens.

SHIU-YING HU
22 Divinity Ave.
Cambridge, Mass., U.S.A.
1976

The Basic Features of
the Orchids in Hong Kong

The orchids constitute the third largest family of flowering plants in Hong Kong. Approximately 78% of the genera are spontaneous and the remaining ones are cultivated exotics. Basic features in morphological diversity and ecological adaptation are discussed under the headings of (1) habit and habitat, (2) root, (3) stem, (4) leaves, (5) inflorescence, (6) flowers, (7) fruit, and (8) seed and seedling.

HABIT AND HABITAT

All the orchids of Hong Kong are perennial herbs owing their existences to partial or complete symbiosis with some fungi. Adaptive features to the subtropical monsoon climate with annual alternation of hot humid and cool dry seasons and to the topographical diversity with the associated ecological conditions are obvious. One-third of the native species pass through the winter dry season in dormancy like the orchids of the temperate region. They disappear after flowering, fruiting, and the dispersal of the seed. *Bletilla, Cheirostylis, Cryptostylis, Disperis, Eulophia, Habenaria, Nervilia, Pachystoma, Pecteilis, Peristylus, Platanthera, Spathoglottis,* and *Zeuxine* are good examples. Species of these genera all have subterranean tubers, corms, or fleshy rhizomes.

1. *Terrestrial orchids*: According to their habitat the native species of orchids in Hong Kong are predominantly terrestrial. Seven-tenths of the genera have the roots in or penetrating the soil, and three-tenths of them are epiphytes with aerial roots. Approximately 71% of the terrestrial species are erect and 29% of them are creeping. There are two types of erect terrestrial orchids, namely the solitary type and the caespitose type. With the solitary type, plants occur scattered in the field, each with a single leafy or flowering-leafy shoot (Fig. 1A). Species of *Brachycorythis, Planthera, Pecteilis, Peristylus, Habenaria,* and *Disperis* are good examples. Plants with this type of growth habit require an annual period of dormancy. At the resumption of physiological activities in the following growing season two phenomena are apparent. In *Disperis, Habenaria, Pecteilis, Peristylus,* and *Platanthera* the corm or tuber gives rise to a leafy flowering shoot. In *Nervilia, Eulophia, Pachystoma,* and *Spathoglottis* the fleshy rhizome produces a flowering shoot and a leafy shoot from two separated buds (Fig. 1B). The caespitose erect terrestrial orchids appear in tufts (Fig. 1C). Plants with this type of growth habit are evergreen. The new growth diverges from the base

of the leafy pseudobulb. *Acanthephippium, Calanthe,* and *Cymbidium* are good examples.

The creeping terrestrial orchids in Hong Kong have slender and rather fleshy rhizomes with elongated smooth internodes and few long roots at the nodes. The annual addition to this rhizome is contributed by an ascending leafy and flowering shoot developed from an axillary bud formed below the apex of the rhizome (Fig. 1D). At the end of the photosynthetic, flowering, and fruiting season, a strong subapical axillary bud develops into a short shoot with a rosette of leaves while the mother ascending shoot falls and becomes an additional part of the rhizome. In the following growing season the short shoot elongates into another ascending flowering shoot. *Anoectochilus, Goodyera, Hetaeria, Ludisia,* and *Vrydagzynea* all have this type of growth habit.

2. *Epiphytes*: Epiphytic orchids have aerial roots attached to the bark of trees, to the surface of wet rocks in shade along the streams, or to exposed boulders subjected to diurnal temperature changes, strong winds, and the xerophytic condition of barren rock. Almost one-half of the epiphytic orchids native to Hong Kong have creeping habit, one-third with trailing habit, and one-fifth with erect caespitose habit.

All the creeping epiphytic orchids of Hong Kong are pseudobulbous. Among them two types of creeping habit are apparent, namely the dichotomous type and the straight-line type. In the dichotomous pattern, two vegetative buds may develop from the base of a pseudobulb simultaneously or at different seasons. Consequently the creeping rhizomes appear branched dichotomously. This pattern is very evident in *Coelogyne* (Fig. 1E). In contrast, in the straight-line pattern only one vegetative bud or a mixed bud is produced annually from the base of a pseudobulb (Fig. 1F). In *Cirrhopetalum, Bulbophyllum,* and *Eria* this bud develops into a short shoot which becomes a rhizome terminated by a leafy pseudobulb. In *Pholidota* this bud develops into a flowering shoot at first and after anthesis it changes into a rhizome with a pseudobulb, and a fruiting cluster between two leaves. In *Bulbophyllum* and *Eria,* the species with straight-line creeping habit may occasionally give rise to dichotomous branching conditions. In other species of these genera such as *B. levinei* and *E. corneri* the internodes of the rhizomes are very short and the pseudobulbs appear crowded (Fig. 1G).

The trailing epiphytic orchids have elongated leafy stems which trial along supports propped and anchored by very long and often branched aerial roots. *Acampe,*

FIGURE 1. Habits and habitats of Hong Kong Orchids: A–D. Terrestrial Orchids. A. The solitary habit of *Platanthera mandarinorum* with a herbaceous deciduous stem terminated by an inflorescence, and a subterranean fleshy tuber. B. The rhizomatous habit of *Eulophia sinensis* showing two types of shoots, the flowering shoot developed earlier in the growing season, and the leafy shoot developed after anthesis. C. The caespitose habit of an evergreen orchid, *Cymbidium ensifolium.* D. The creeping habit of *Hetaeria nitida* showing the terminal inflorescence and the slender rhizome consisting of five years' growth, each developed from a subterminal bud. E–I. Epiphytic orchids. E. The dichotomous branching of the rhizome of a pseudobulbous species, *Coelogyne fimbriata,* showing a flowering shoot developed at the apex of a mature pseudobulb. F. The straight-line type of rhizome of *Pholidota cantonensis,* showing the pseudobulbs of three years' growth, and current year's flowering shoot, one of the internodes is capable of developing into a pseudobulb with one or two leaves after anthesis. G. The habit of *Eria corneri* showing the crowded pseudobulbs. H. A monopodial epiphytic species, *Diploprora championii,* showing the elongated aerial roots, lateral inflorescences and leafy stem with continuous apical growth. I. A sympodial caespitose orchid, *Appendicula bifaria,* showing five years' growth, the distichous leaves and the axillary inflorescences on the fourth year's growth.

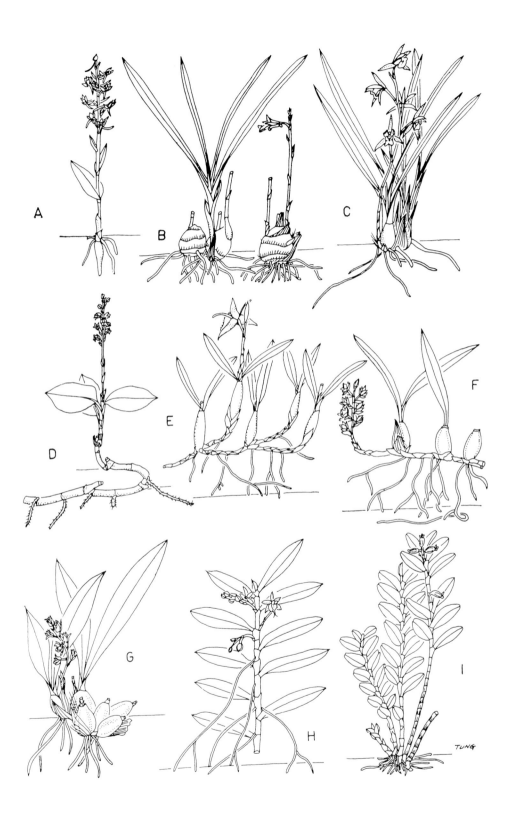

3

Cleisostoma, Diploprora, Robiquetia, Thrixspermum, and *Vanda* are good examples (Fig. 1H).

Appendicula bifaria and *Dendrobium acinaciforme* are two species of native epiphytic orchids with caespitose growth habit. In these species the new shoots emerge from the bases of the older ones (Fig. 1I). In *Appendicula* the clump may attain a diameter of 25 cm. and the plant has a cushion-like appearance.

3. *Symbiosis*: Symbiotic relationship with a fungus has obvious effects on the habit of some orchids in Hong Kong. *Eulophia yushuiana* is a saprophyte. It depends completely upon fungal supply of food as the energy source and of minerals. The plant consists of a large compressed oblong potato-like rhizome with concentric rings. Its aerial structure is adapted for reproduction, i.e., the flowering and fruiting scape.

Cheirostylis is an epiphyte growing on wet boulders along streams in forested areas. It is associated with unicellular or filamentous algae, some leafy hepatophytes and occasionally with small colonies of low mosses. A young plant has a worm-like succulent purplish olive-brown rhizome or small leaves of similar color at one end. In a major portion of the year the plant is leafless. On the contact-side of the rhizome there are numerous soft hairs which adhere closely to the substratum. Structures of this nature are infiltrated by fungi which are responsible for the decomposition of the debris of the associated cryptogams on the rock and for the absorption of food and minerals. When the plant is old enough to flower, a scape terminal to the rhizome with 1–4 leaves is produced (Fig. 16).

The growth habit of *Zeuxine strateumatica* is worthy of note. The species has a very short life span above ground. The plant body consists of a tender erect leafy and flowering shoot which is the continuation of a delicate subterranean rhizome bearing few small brown tubers. The leafy and flowering shoot appears rather inconspicuously among grasses, and completes the reproductive processes within two months. Then it disappears completely from the surface of the soil. Apparently the seedlings and the tubers develop and increase in size underground through symbiosis with a beneficial fungus. This is a subject worthy of further investigation.

ROOT

Anchorage is the primary function of root in orchids, especially with the epiphytic species. In epiphytic orchids with short internodes, such as *Gastrochilus* and *Phalaenopsis*, numerous roots crowd below the leaves and adhere closely to the substratum to fasten the plants firmly in position. The epiphytes with long internodes, such as *Acampe, Arachnis, Cleisostoma, Diploprora, Thrixspermum,* and *Vanda*, produce roots among leaves. These first extend to the supporter, prop the plant in position, and then branch and adhere to the substratum. The pseudobulbous epiphytes, such as *Bulbophyllum, Coelogyne, Eria,* and *Pholidota,* normally produce roots below the pseudobulbs.

The aerial root of epiphytic orchids is silvery gray, except 2 cm. from the root tip which is green. This color is due to the presence of a spongy multicellular epidermal layer of dead cells, the velamen. This protective layer absorbs moisture readily. Some orchidologists maintain that the velamen absorbs dew and rain and conserves the water temporarily for use in the period of drought.

Some aerial roots are greenish, especially near the apical portion. The chlorophyll of the roots enables the plants to manufacture food. In aphyllous species such as *Microcoelia guyoniana,* the roots are the only photosynthetic organs.

The fleshy roots of many terrestrial orchids that require dormancy, such as

Cryptostylis and *Spiranthes*, are the storage organs that conserve food and water during the dry season. The feeding roots of orchids enable the plants to obtain nutrient supplies from the beneficial symbiotic fungi.

STEM

The stems of orchids in Hong Kong display a high degree of morphological diversity. In the tribe Orchideae, the aerial stems are erect and deciduous. A stem of the mature plant emerges from a tuber or a corm. It bears leaves and is terminated by a raceme or a spike. After completing the production of flowers, fruits, and seeds, it dies and disintegrates, leaving a corm or tuber to continue the life cycle. *Brachycorythis, Disperis, Habenaria, Pecteilis, Peristylus,* and *Platanthera* all have this type of stem.

In the tribe Neottieae, the stems persists as rootstocks or rhizomes for several years. In *Cryptostylis, Manniella,* and *Spiranthes,* the rootstocks are very short and erect. Two buds are produced by *Cryptostylis* and *Manniella,* one of which develops into a flowering shoot and the other to a leafy shoot. In *Manniella* the flowering shoot appears much earlier than the leafy one. In *Spiranthes* the short rootstock produces 2 or 3 buds, with one or two developing into a rosette of leaves with a scape at the center. In *Anoectochilus, Hetaeria, Ludisia, Vrydagzynea* and several species of *Goodyera* the stems are slender and with long internodes. The stem of a mature plant of these genera is developed from an axillary bud near the apex of a rhizome. It is leafy and is terminated by a raceme. It also produces a strong axillary bud for repeating the vegetative and reproductive functions in the following year.

In the tribe Epidendreae the stems may have continuous apical growth (monopodial) or they may have arrested apical growth (sympodial). One-third of the 44 genera of this tribe growing in Hong Kong have monopodial stems. The species are all epiphytic. The length of the stems varies from 1 cm. in *Microcoelia* to 2 m. or more in *Vanda* and *Arachnis*. They all produce lateral flower buds.

The other two-thirds of the genera of this tribe in Hong Kong have sympodial stems. These may be erect or creeping. The species with erect sympodial stems are epiphytic. The length of the stems varies from an average of 20 cm. in *Appendicula* to approximately 1 m. in a variety of *Arundina chinensis* and some species of *Dendrobium*. These stems may produce terminal inflorescences such as in *Epidendrum* and *Arundina,* or they may have lateral flower buds such as in *Appendicula* and *Dendrobium*.

The creeping epiphytic stems are essentially rhizomatous. The rhizomes may be slender and long, or stout and short. One or more of the internodes of a rhizome may be specialized for the conservation of food and water, and become enlarged and fleshy. This enlarged fleshy internode or internodes is called a pseudobulb. The rhizome or pseudobulbous rhizome of a mature plant may be terminated by a flowering branch such as in *Liparis* and *Pholidota*. It may produce a basal-lateral flower bud which develops into a scape, such as in *Bulbophyllum* and *Cirrhopetalum*. In *Coelogyne,* a pseudobulbous rhizome produces a terminal flower bud which develops into a scape several months after the leaves are fully expanded.

The types of the stems determine the habit of the species. They are of fundamental importance in the classification of the orchids, in the propagation of the species and in the economic importance to man. The pseudobulbs of *Pholidota* and the stems of several species of *Dendrobium* are used as medicine.

LEAVES

The arrangement, size, shape, and texture of the leaves of orchids in Hong Kong reflect their habits, habitats, and taxonomic positions. The terrestrial orchids of the tribes Cypripedieae, Orchideae and Neottieae have tender, herbaceous, small or medium-sized, and basal or cauline leaves without coarse veins. The leaves of the tribe Epidendreae are more diversified morphologically. The terrestrial species in *Acanthephippium, Calanthe, Cephalantheropsis, Geodorum, Malaxis, Pachystoma, Phaius,* and *Spathoglottis* have large, thin, papery and plicate-venose leaves. The erect epiphytic orchids such as *Appendicula, Arundina,* and *Dendrobium* have numerous, leathery, small or medium-sized leaves distichously arranged on the stem. The creeping pseudobulbous epiphytic orchids such as *Coelogyne, Bulbophyllum, Cirrho-petalum, Eria,* and *Thelasis* have 1–4 leaves apical to the pseudobulb. The trailing monopodial epiphytic orchids such as *Acampe, Arachnis, Cleisostoma, Diploprora, Thrixspermum,* and *Vanda* have many medium-sized or large band-like leathery and rather fleshy leaves. Those of *Cleisostoma teres* and *Vanda teres* are cylindrical.

The leaves of the epiphytic species are modified for the conservation of water. The laminas are covered by waxy cuticle. They contain a considerable amount of storage tissue. The petioles are tubular, tightly ensheathing the internodes of the stem. Between the lamina and the sheathing portion of the petiole there is an articulation where the lamina falls. The petiole is persistent and it becomes straw-colored after the fall of the lamina.

The leaves of *Cymbidium* are grass-like, and those of *Mischobulbum* and *Nephelaphyllum* are ovate or cordate, rather fleshy and on extremely short petioles. The leaves of an exotic African species, *Microcoelia guyoniana*, cultivated in Hong Kong, are reduced to imbricate scales. The species is aphyllous. The spontaneous saprophytic species, *Eulophia yushiana* and *Aphyllorchis montana*, have no leaves.

INFLORESCENCE

The flowers of the orchids in Hong Kong appear to be in clusters opening in acropetal succession. Actually, there are many evidences indicating that some of the clusters represent monochasial cymes. The zigzag rachis of *Pholidota*, the spiral arrangement of the flowers of *Spiranthes* and *Peristylus*, and the delayed development of the succeeding flowers of *Ceologyne* and *Arundina* are good examples. For the convenience of the readers, here conventional terms used in classical references are adopted for describing the arrangement of flowers in the cluster (inflorescence). They are (1) raceme for a simple elongated cluster with pedicellate flowers; (2) spike for a similar cluster with sessile flowers; (3) umbel for a cluster with very short or sometime hardly visible rachis; and (4) panicle for a cluster with long and branched rachis. Solitary flowers may occur occasionally in case of extreme reduction and abortion of flower bud, such as in *Bulbophyllum ambrosia*.

The position of an inflorescence has been taken by many orchidologists as a criterion in the classification of orchids. Accordingly, the tribe Epidendreae is roughly subdivided into two series, the Acranthae with terminal inflorescences and the Pleuranthae with lateral inflorescence. Examples for Series Acranthae are *Arundina, Coelogyne, Liparis, Malaxis, Pholidota,* and *Epidendrum*, and those for Series Pleuranthae are *Calanthe, Phaius,* and all the monopodial species.

FLOWERS

Morphological diversity and structural complexity culminate in the flowers of orchids. During a period of one and one-half centuries in the past, orchidologists have designed special terms for describing the floral structure of orchids and such terminology is of international significance. For the convenience of the readers, an illustrated account on the (1) basic structure and common phenomenon, (2) sepals, (3) petals, (4) lip, (5) column, (6) anther and pollinia, and (7) ovary and stigma is provided here.

1. *Basic Structure and Common Phenomenon*:   An orchid flower is basically trimerous, pentacyclic and bilaterally symmetrical. Modifications of this basic structure occur in the change of the segments and through the reduction and connation of parts.

The trimerous condition is most obvious with the outermost whorl, the sepals. Segments of the next whorl are differentiated into two petals and a lip. Reduction is most eminent with the staminal whorls. In the diandrous orchids of the subfamily Cypripedioideae, two of the stamens of the outer staminal whorl are reduced and the third one is modified into a butterfly-like staminode (Fig. 2A, C), while two anthers of the inner staminal whorl become functional. In Hong Kong only *Paphiopedilum* has this basic structural pattern. In the monandrous orchids of the subfamily Orchidoideae two of the anthers of the outer whorl are reduced and the third one becomes the functional anther of the flower while the anthers of the inner staminal whorl are all reduced (Fig. 2B). All the genera of orchids in Hong Kong except *Paphiopedilum* have this pattern of structure. The union of parts occurs between the androecium and gynoecium where the filaments and the style connate and adnate to form the column. In the diandrous *Paphiopedilum* the discoid stigma is on a stalk and the anthers are lateral (Fig. 2C). In the monandrous orchids, the anther is terminal to the column while the stigma (or stigmas) is in a lower position. These two organs are separated by a rostellum (Fig. 2D). The ovary of orchid is inferior, and at anthesis it is not tumid.

A common phenomenon of the twisting of the ovary and/or the pedicel (resupination) for 180° occurs with 88% of the native orchids of Hong Kong. The twist occurs with a mature flower bud just before it opens into a flower (Fig. 2E). Only in *Appendicula, Cryptostylis, Hetaeria, Malaxis, Nephelaphyllum,* and *Thelasis* are the flowers non-resupinate.

2. *Sepals*:   The sepals of most orchids are subequal, free, and spreading. For convenience, many orchidologists call the sepal opposite the lip the odd sepal, and the remaining two the lateral sepals. In a resupinate orchid flower the odd sepal is in a superior position, and in a non-resupinate flower it is inferior.

Certain amount of union does occur with the sepals. In *Acanthephippium, Appendicula, Bulbophyllum, Cirrhopetalum, Dendrobium, Eria,* and *Mischobulbum,* the lower margins of the basal portion of the lateral sepals adnate to the column foot to form a mentum (chin). In *Cirrhopetalum* the same sepals twist beyond the mentum with their upper margins meeting before the lip and forming a coherent platform 2–6 times longer than the odd sepal of the same flower (Fig. 2I). In *Cheirostylis* all the sepals are connate to form a tube enclosing the petals and the column (Fig. 2G). In *Manniella* the basal portion of the sepals are connate and adnate to the upper portion of the

7

ovary to form a pouch enclosing the appendages of the lip (Fig. 2H). The lateral sepals of *Phaphiopedilum* are united into one sepal-like structure (the synsepalum) normally hidden by the slipper-like lip.

3. *Petals*: The orchids of Hong Kong generally have petals subequal or slightly smaller than the sepals. Approximately 70% of the genera have spreading petals. *Arundina, Cymbidium, Dendrobium, Epidendrum, Phaius,* and *Vanda* are good examples. The remaining 30% of the genera have small petals, situated below and connivent with the odd sepals. Together they form a hood over the column of the flower. *Goodyera, Habenaria, Peristylus,* and *Platanthera* are good examples. The petals of *Pecteilis susannae* and *Coelogyne fimbriata* are linear or filiform, and spreading.

4. *Lip (Labellum)*: The lip of an orchid flower is the highly specialized floral part. The shape, division, attachment, and extraordinary growth (processes) of the lip are important criteria for distinguishing the species. The orchids of Hong Kong display two basic types of lip, i.e., the simple undivided lip and the 3-lobed lip.

FIGURE 2. Basic structure of the flowers of orchids in Hong Kong: A–D. Two structural patterns of orchid flowers. A. A diagram of the floral parts of a diandrous orchid, *Paphiopedilum purpuratum*, showing the positions of the odd and the connate lateral sepals, the petals and the lip, the staminode and the reduced stamens in the outer staminal series, the functional stamens and of the theoretical reduced stamen in the inner staminal series and the tricarpellate ovary. B. A diagram of the floral parts of a monandrous orchid, *Habenaria dentata,* showing the positions of the distinct sepals, the petals, the lip, the functional and the theoretical reduced stamens of the outer staminal series, the glands and the reduced stamen of the inner staminal series, and the tricarpellate ovary. C. A side view of the apical portion of column of a diandrous orchid, showing the staminode, one lateral functional anther and the discoid stigma on a stalk. D. The front view of an ovary and column of a monandrous orchid (LEFT) and the longitudinal section of the same (RIGHT) showing the positions of the anther, stigma and ovules. E. A portion of the raceme of *Phaius tankervilliae*, showing the common phenomenon of resupination as indicated by the position of the spur of the buds and of an open flower. F–I. Unusual shape and structure of sepals as displayed by Hong Kong orchids. F. The lateral view of a flower of *Bulbophyllum levinei* showing the partial union of the bases of the lateral sepals with the column foot to form a mentum. G. The lateral view of a flower of *Cheirostylis chinensis*, showing the tubular connate sepals. H. The lateral view of a flower of *Manniella hongkongensis*, with portion of the odd sepal, a lateral sepal and a petal removed, showing the sepals connate and adnate to the apical portion of the ovary forming a pouch below the column. I. The top view of a flower of *Cirrhopetalum tigridum*, showing the twisted lateral sepals with the upper margin connivent in front of the lip forming a platform. J–Y. The lips of orchid. J. The undivided concave lip of *Pholidota cantonensis*. K. The flat undivided lip of *Cryptostylis arachnites*. L. The undivided and calcarate lip of *Platanthera angustata*. M. The twisted lip of *Ludisia discolor*. N. The auriculate lip of *Malaxis latifolia*. O. The trilobed lip of *Spathoglottis pubescens*. P. The saccate lip of *Thrixspermum centipeda* with erect lateral lobes. Q. The calcarate trilobed lip of *Pecteilis susannae*, showing the large fimbriate lateral lobes and the smaller entire midlobe. R. The lip of *Liparis plicata* showing calli at the base. S. The lip of *Bletilla striata* showing the lamellate disc and midlobe. T. The lip of *Pachystoma chinense* showing the warty and papillose disc. U. The lip of *Oncidium varicosum* showing the crested calli. V. The lateral view of a flower of *Vrydagzynea nuda* with portions of odd sepal and of the lip, a petal and a sepal removed, showing calli in the spur. W. The lateral view of a flower of *Cleisostoma fordii* with sepals, petals and portion of lip removed, showing the callus in the back of the spur produced forward and becoming connivent with the keel which extends from the disc downward to the base of the spur. X. The column and one-half of the lip of *Cymbidium ensifolium* showing a keel on the disc. Y. The lateral view of a flower of *Calanthe patsinensis* with sepals and petals removed, showing the sides of the lip adnate to the column.

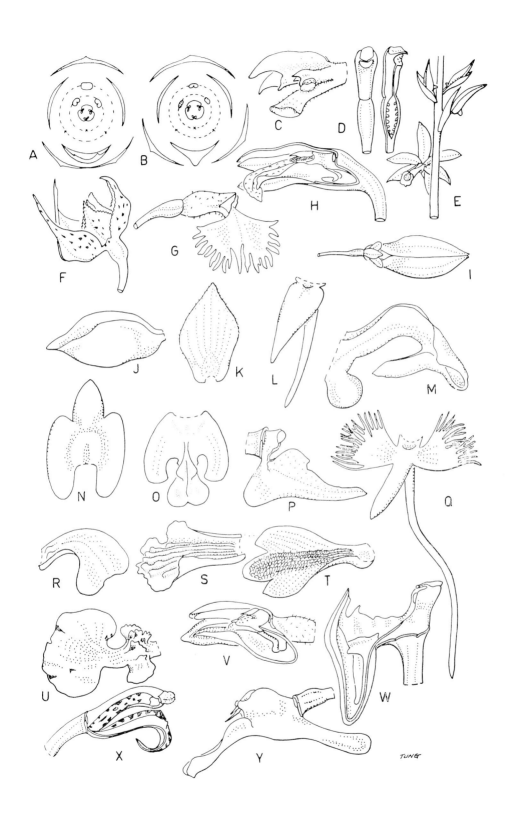

Many genera of Hong Kong orchids have undivided lips. *Cryptostylis, Geodorum, Goodyera, Hetaeria, Liparis, Malaxis, Platanthera,* and *Thelasis* are good examples. Most of the undivided lips are concave (Fig. 2ᴊ), and those of *Goodyera* and *Hetaeria* are ventricose. In *Liparis* the lips are tongue-like and reflexed (Fig. 2ʀ) and in *Ludisia* the lips are twisted. In *Cryptostylis* (Fig. 2ᴋ) and some species of *Malaxis* the lips are flat. In *Malaxis* the lips are auriculate (Fig. 2ɴ). The undivided lips of *Platanthera* are spurred (calcarate, Fig. 2ʟ).

The lip of approximately 50% of the genera of Hong Kong orchids is distinctly 3-lobed. The space between the lobes is the disc. The lateral lobes (side-lobes) may be erect (Fig. 37, *Bletilla*), or spreading (Fig. 6ᴀ, *Pecteilis*). The midlobes may be relatively small and tongue-like (Fig. 2ǫ, *Pecteilis*), enlarged and spreading (Fig. 2ᴜ, *Oncidium*), or keeled and reflexed (Fig. 2x, *Cymbidium*). The lips of *Phaius, Arundina,* and *Cattleya* are trumpet-shaped, embracing the column, without distinct side-lobes, and wavy, crisped, recurved at the tip (Fig. 51, *Phaius*). The lips of *Paphiopedilum* is slipper-like (Fig. 5ᴀ).

The lips of orchids are generally broadly attached to the base of the column, and they are immovable. In *Bulbophyllum* and *Cirrhopetalum* the fleshy tongue-like lips are hinged to the end of the column foot, and they are exceedingly mobile (Fig. 3ᴊ). The lip of *Calanthe* or *Epidendrum* is adnate to the column forming a pouch below the stigma (Fig. 2ʏ, *Calanthe*).

The lips of many orchids are decorated with unusual growths. These may be calli on the base (Fig. 2ʀ, *Liparis*), keels on the disc (Fig. 2x, *Cymbidium*), lamellate midlobes and discs (Fig. 2s, *Bletilla*), crested calli (Fig. 2ᴜ, *Oncidium*), papillose warts on disc (Fig. 2ᴛ, *Pachystoma*), calli in the spurs (Fig. 2v, *Vrydagzynea*), and basal appendages (Fig. 2ʜ, *Manniella*). In *Cleisostoma fordii,* the callus on the back of the spur extends forward and becomes connivent with the keel which extends down from the disc (Fig. 2w).

5. *Column*: Morphologically the column of an orchid is the stalk which bears the pollination apparatus. The size, shape, and structure of the column have direct bearings on the anther and the stigma. In connection with the basic structure of orchid flowers mention is made of the column of subfamily Cypripedioideae. The discussion here refers to the columns in subfamily Orchidoideae only.

The orchids of Hong Kong in subfamily Orchidoideae display two basic types of column, i.e., the column bearing two separated thecae on the side (Fig. 3ᴀ, ʙ), and the column with a terminal anther (Fig. 3ᴅ). All the genera in the tribe Orchideae have columns bearing two separate thecae on the side. These columns are short, stout, and with two lateral glands near the pointed lower ends of the thecae. Variations of this type of columns accompany the number and position of the stigma or stigmas. In *Brachycorythis* and *Platanthera* where there is one stigma, the rostellum on the column is narrow and with short arms (Fig. 3ᴀ). In *Habenaria* and *Pecteilis* where there are two stigmas, the rostellum is broad and with long arms (Fig. 3ʙ, *Habenaria dentata*).

The columns of orchids in the tribes Neottieae and Epidendreae are characterized by having a terminal anther. They display a wide range of morphological variations. There are columns long and undecorated (Fig. 3ᴅ, *Cymbidium*), short and with glands (Fig. 3c, *Cryptostylis*), short and with front appendages (Fig. 3ᴇ, *Hetaeria*), straight and broadly winged (Fig. 3ꜰ, *Cirrhopetalum*), simple and twisted (Fig. 3ɢ, *Ludisia*), highly decorated with colored trichomes (Fig. 3ʜ, *Cleisostoma*), enlarged and petaloid

(Fig. 3I, *Oncidium*), armed at the apex (Fig. 3J, *Bulbophyllum*), narrowed at the base (Fig. 3D, *Cymbidium*), and produced into a long foot at the base (Fig. 3J, *Bulbophyllum*).

On the column, between the anther (or anthers) and the stigma (or stigmas), there is a delicate microscopic organ, the rostellum. The orchids of Hong Kong display three types of rostellum. Those of *Brachycorythis, Disperis, Habenaria, Pecteilis Peristylus,* and *Platanthera* are bent, yoke-shaped, extending laterally into slender arms, and on the ends of which rest the viscidia (Fig. 3A, *Brachycorythis*; and B, *Habenaria*).

Approximately 35% of the genera of Hong Kong orchids have truncate or subtruncate rostella (Fig. 3D, *Cymbidium*; and F, *Cirrhopetalum*). The truncate rostellum of *Dendrobium* is broad, slightly constricted above the base and recurved along the upper margin (Fig. 3K). In *Pholidota* the rostellum rises above the plane surface of the column and bends over the stigmatic cavity (Fig. 3L). The rostellum of *Cleisostoma teres* is horse-shoe-shaped (Fig. 3H).

Almost 35% of the orchid genera in Hong Kong have beaked or subulate and generally bifurcate rostella (Fig. 3G, *Ludisia*; M, *Goodyera*; and O, *Manniella*). In *Ludisia* the rostellum is fleshy, twisted, and curved. It supports the viscidium on the short side of the curve. In *Calanthe patsinensis* the rostellum is triangular-ovate, acumin-ate-cuspidate at the apex, and extending a flexible slender portion into the sheath of the linear viscidium (Fig. 3N). The bifurcate rostella of *Cheirostylis, Calanthe, Goodyera, Hetaeria, Vrydagzynea,* and *Zeuxine* are firm, and each supports a viscidium between the equal lobes. In *Manniella*, the arms are unequal, and each lobe has a discoid viscidium at the apex (Fig. 3o).

There are a few unusual cases worthy of note. The rostellum of *Spiranthes sinensis* is very delicate, white, hyaline, pointed, and bifid at the apex. The rostellum of a closely related but hairy species, *S. hongkongensis*, is similar in texture but very much reduced, and attached to the middle of the pollinia (Fig. 14D). In *Cryptostylis* the column is very short, and the viscidium is in direct contact with the stigma. No rostellum is evident (Fig. 3D).

6. *Anther and Pollinia (Pollinium)*: The number and position of the anther (or anthers), and the texture, attachment and number of the pollinia are of fundamental importance in the classification of orchids. The major groups of the family, i.e., the subfamilies, divisions, tribes, subtribes and often the genera are defined primarily on the characters of the anthers and the pollinia. The subfamily Cypripedioideae is characterized by two fertile anthers on the column opposite the petals and a staminode above the stigma (Fig. 2A, C). In contrast the subfamily Orchidoideae is characterized by one fertile anther terminal to the column above the stigma opposite the lip (Fig. 2B, D). The two divisions in Orchidoideae are divided on the basis of the attachment of the anther on the column. The division Basitonae is distinguished by erect anther with widely separated thecae, each closely adnate to the front of the column by a broad base (Fig. 3A, B), and by two sectile pollinia each with a slender caudicle terminated by a viscidium (Fig. 4A). In contrast the division Acrotonae is distinguished by an incumbent or reclining (rarely erect) anther attached to the column by a short and thin, often filiform filament, and by 2–8 pollinia diversified in texture, shape, and pollination mechanism. The tribes in Acrotonae are distin-guished primarily by the texture of the pollinia and the stability of the anther caps. Tribe Neottieae is characterized by soft granular or sectile pollinia and the anther

caps persistent after pollination, while the tribe Epidendreae is characterized by waxy or cartilaginous pollinia and deciduous anther caps. Variations in the shape of the anther correspond to the differences in the shape of the column, the structure of the rostellum and the pollinia. Many monandrous genera of Hong Kong orchids have oblong, ovoid, hemispherical or globose anthers rounded, truncate or obtuse at the apex (Fig. 4c–h). In *Dendrobium* the spherical anther is evidently lobed (Fig. 4c). In *Acanthephippium* the oblong anther is helmet-like (Fig. 4e). In *Eulophia* the ovoid anther is produced at the apex and with 2 round lobes (Fig. 4f). In *Nephelaphyllum* and *Tainia* the anthers are bicornute (Fig. 4h). Many other genera of Hong Kong orchids have ovate-lanceolate anthers acuminate or beaked at the apex (Figs. 3m,n; 4b). Anthers of different shapes may occur with species of the same genus. For example, in *Cleisostoma*, the anther of *C. teres* is truncate and that of *C. fordii* is beaked.

Pollinia are small yellow microscopic masses of agglutinated pollen grains. The orchids of Hong Kong display a dramatic exhibition of the forms and attachments of pollinia. Many genera have 8 pollinia. These pollinia may have slender caudicles attached to a viscidium (Fig. 4k, *Cephalantheropsis*), slender stipe with a viscidium (Fig. 4l, *Thelasis*), or they may have no caudicle, stipe, or evident viscidium (Fig. 4i, j, m). They may be equal, obovoid (Fig. 4i, *Mischobulbum*), or compressed and discoid (Fig. 4j, *Tainia*). They may be unequal (Fig. 4m, *Phaius*). Many genera have four pollinia, and *Bulbophyllum, Cattleya, Cleisostoma, Goodyera, Epidendrum, Robiquetia,* and

---

FIGURE 3. The column, anther, rostellum and stigma of Hong Kong orchids: A–B. Short and stout columns with the thecae attached by the broad bases to the front and with 2 glandular auricles. A. Column of *Brachycorythis galeandra* with 1 stigma. B. Column of *Habenaria dentata* with 2 stigmas. C. Very short column of *Cryptostylis arachnites* with the erect anther slightly pushed back, showing the glandular appendage, the large stigma, and the apical viscidium. D. The simple and long column of *Cymbidium ensifolium*, showing the subglobose anther and the stigma. E. The lateral view of a flower of *Hetaeria cristata* with the sepals, petals and portion of the lip removed, showing the front appendage of the column, lateral gland and the persistant erect anther cap. F. The winged column of *Cirrhopetalum tseanum*, showing the hemispherical anther, the stigmatic cavity with a large glands below the truncate rostellum and 2 smaller glands at the opposite margin. G. The twisted column and rostellum of *Ludisia discolor* showing the viscidium in the concave side of the oblique rostellum and the stigma on a plain surface. H. The front view of the column of *Cleisostoma teres*, showing the trichomes and the horse-shoe-shaped rostellum. I. The petaloid column of *Oncidium varicosum*. J. The lateral view of a flower of *Bulbophyllum youngsayeanum* with portion of the odd sepal, a lateral sepal and a petal removed showing the column with two arms at the apex, an extended foot at the base, and the movable lip. K. The apical portion of the column of *Dendrobium acinaciforme* showing the broad rostellum reflexed along the upper margin. L. The column of *Pholidota chinensis* with the terminal anther slightly pushed back, showing the pollinia attached to a viscidium at the apex of the anther, and the rostellum rising and bending over the stigmatic cavity. M. The column of *Goodyera youngsayei*, showing the beak-like bifid rostellum and the discoid stigma on the truncate end of the column. N. The apical portion of the column of *Calanthe patsinensis* showing the ovate-acuminate anther cap, the triangular-ovate rostellum with a flexible slender portion extending to the sheath of the linear viscidium. O. The front view of the apical portion of the column of *Manniella hongkongensis* showing the rostellum with unequal lobes each with a viscidium, and two stigmas. P. The front view of the column of *Calanthe triplicata* with the lip partially removed, showing the stigmas exposed on the apex of the column and connivent below the bifid rostellum. Q. The front view of the column of *Calanthe striata* with the lip partially removed, showing the U-shaped stigma completely hidden by the rostellum.

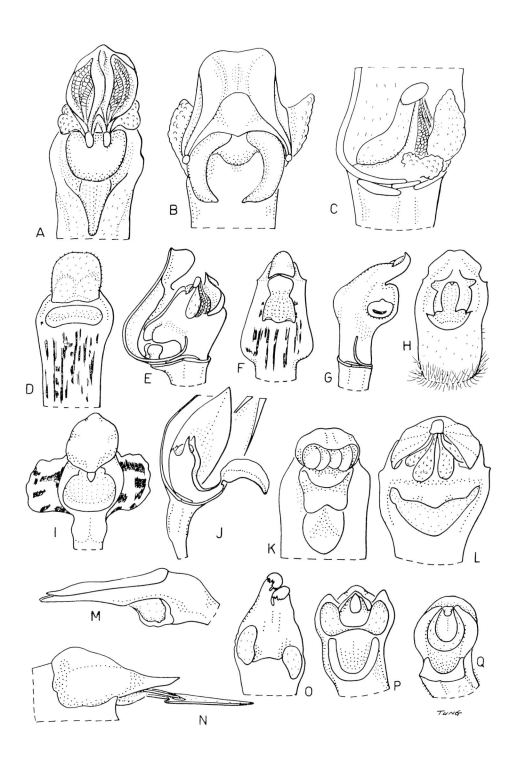

Thrixspermum are good examples. In *Bulbophyllum* the pollinia are unequal, and they are not associated with obvious viscidium. In *Cattleya* the pollinia are compressed, equal and each pair is attached loosely on granular appendage (Fig. 4o). In *Epidendrum* the two pairs of pollinia are attached to a granular stipe-like structure with a viscidium on the back (Fig. 4p). In *Thrixspermum* the pollinia are unequal and the pairs are attached to a white viscidium with a recurved ovate membrane (Fig. 4q). In *Robiquetia* the pollinia are slightly unequal and the pairs are attached on a slender and elongated stipe with a viscidium (Fig. 4r). In *Cleisostoma teres* the two pairs of pollinia are attached to a short stipe with broad horse-shoe-shaped base (Fig. 4s). In *Goodyera procera* the 4 pollinia are in two pairs attached to a viscidium. In other species of the genus the pollinia may have slender caudicles attached to a viscidium.

Many genera of Hong Kong orchids have two waxy pollinia on stipes each terminated by a viscidium. These pollinia appear subglobose in front. Actually under a microscope with good illumination they are seen to be folded, cleft or concave on one side. In *Phalaenopsis* the pollinia are obovoid and hollow on the back (Fig. 4u), while in *Vanda* they are subglobose and cleft on the back (Fig. 4v). In *Gastrochilus* the pollinia are hemispherical in appearance and concave on the back, and the viscidium is bifid (Fig. 4w). *Appendicula* is the only genus of Hong Kong orchids which has six pollinia on two slender caudicles (Fig. 4x).

7. *Ovary and Stigma (or Stigmas)*:   The ovary and the stigma represent the gynoecium of an orchid flower. As an inferior ovary is a family characteristic, the ovary of an orchid flower is below the sepals and the petals. At the flowering stage the ovary is hardly distinguishable from the pedicel. Under the microscope, the ovary appears

---

FIGURE 4.   The anthers and pollinia displayed by Hong Kong orchids: A. A sectile pollinium of *Peristylus spiranthes* with a slender caudicle and a viscidium.   B–H. Shape and attachment of anthers.   B. A top view of the column of *Zeuxine gracilis*, showing the ovate anther acuminate at the apex.   C. The back view of the anther of *Dendrobium hercoglossum*, showing the attachment by a short filament.   D. The front view of the anther and pollinia of *Liparis odorata* with the cap pushed up, showing 2 anther cells and the attachment by a thin membrane.   E. The front view of a helmet-like anther of *Acanthephippium sinense*.   F. The front view of the ovoid anther of *Eulophia sinensis* bilobed at the top.   G. The globular anther of *Liparis macrantha*.   H. The bicornute anther of *Tainia dunnii*.   I–M. Eight pollinia and variations in shape and attachment.   I. Obovoid equal pollinia of *Mischobulbum cordifolium* without caudicle and viscidium.   J. The discoid pollinia of *Tainia dunnii* without caudicle and viscidium.   K. Obovoid pollinia of *Cephalantheropsis gracilis* with slender caudicles and viscidium.   L. Small obovoid pollinia of *Thelasis hongkongensis* on elongated stipe with viscidium.   M. Unequal pollinia of *Phaius tankervilliae*.   N–T. Four pollinia and variations in shape and attachment.   N. Unequal pollinia of *Bulbophyllum radiatum* without caudicle and viscidium.   O. Equal and strongly compressed pollinia of *Cattleya lueddemanniana* loosely attached on granular appendage.   P. The pollinia of *Epidendrum ibaguense* on a granular stipe-like structure with a viscidium on the back.   Q. The unequal pollinia of *Thrixspermum centipeda* on membraneous stipe with small viscidium.   R. The pollinia of *Robiquetia succisa* on an elongated and curved stipe with viscidium.   s. The front and back views of the pollinia of *Cleisostoma teres* on a stipe horse-shoe-shaped at the base.   T. The sectile pollinia of *Goodyera procera* with viscidium.   U–W. Two pollinia and variations in shape and attachment, showing both the front and back views.   U. The subglobose-obovoid pollinia of *Phalaenopsis amabilis*, hollow on the back and on spathulate stipe with large viscidium.   V. The oblong-ovoid pollinia of *Vanda teres* deeply cleft on the back and on broad stipe with a large and recurved viscidium.   W. The hemispherical pollinia of *Gastrochilus holttumianus* concave on the back and on a slender stipe with a bifid viscidium.   x. Six pollinia of *Appendicula bifaria* in two groups, and on slender caudicles with a viscidium.

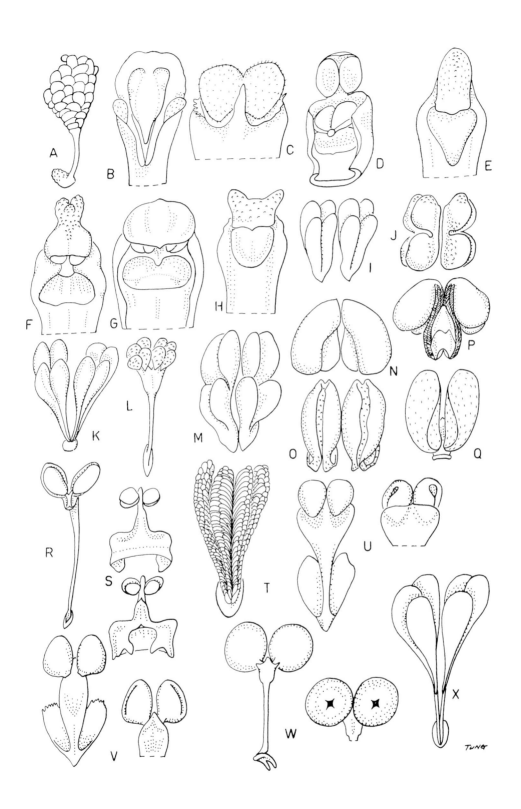

more striate-sulcate than the pedicel. Due to the resupination of the flower, the ridges and grooves are often oblique. The ovaries of a few genera of Hong Kong orchids are very hairy. *Anoectochilus*, *Pachystoma*, and *Spathoglottis* are good examples.

The stigma (or stigmas) of an orchid flower is situated on the apical or front portion of the column. In regard to number, approximately 10% of the genera of Hong Kong orchids have 2 stigmas (Fig. 3B, O), and the remaining 90% of them have one stigma (Fig. 3A, C, D, F–I, K–M, P–Q). Regarding the position of the stigma (or stigmas), it may be on a plane surface of the column (Fig. 3C, *Cryptostylis*; G, *Ludisia*; and O, *Manniella*), in a cavity of the column (Fig. 3F, *Cirrhopetalum*; I, *Oncidium*; K, *Dendrobium*; and L, *Pholidota*), on a truncate end of the column (Fig. 3M, *Goodyera*), on the sides of the apex of the column and connivent below the rostellum (Fig. 3P, *Calanthe triplicata*) or completely hidden below the rostellum (Fig. 3Q, *Calanthe striata*).

FRUIT

The fruit of an orchid is a capsule developed from a tricarpellate inferior ovary. Capsules of 23 genera are available for my studies. Characters common to all the genera observed are dehiscence by six valves, parietal placentation, and the differentiation of spongy and coriaceous tissues responsible for the hygroscopic sensitivity of the valves.

At the maturity of a capsule the six valves separate with the ends connate. The outer three valves are relatively narrow and firm. They are linear and parallel along the margin. The inner three valves are broader and thinner than the outer ones. They are more or less elliptic, each with a median longitudinal ridge on the outside, and becoming progressively thinner toward the margin. Each of the inner valves has a median longitudinal placenta in line with the outer ridge. In *Pholidota chinensis* where the outer ridges of the capsule are high, the placentas appear to be in channels on the inside of the valves. Under a dissecting microscope, the tissues of the valves appear spongy without and coriaceous or thinly cartilaginous within. The valves are hygroscopically sensitive.

Variations in the shape, size, polarity, and appendage of the capsules are obvious. The longest capsule observed is that of *Acampe* which is 7 cm. long. The shortest capsules are observed in *Goodyera procera*, *Liparis longipes*, and *Malaxis parvissima* where they are 4 mm. long. Capsules of *Appendicula*, *Spiranthes*, and *Zeuxine* are slightly longer, and they are 6–8 mm. long. The capsules intermediate in size are found in *Cleisostoma*, *Coelogyne*, *Cymbidium*, *Diploprora*, and *Pholidota*. These vary from 1.5 cm. to 4 cm. in length. The long capsule of *Acampe* is cylindric. The short capsules of *Appendicula*, *Spiranthes*, and *Zeuxine* are oblong or obovoid. The capsules of *Cleisostoma*, *Diploprora*, and *Cymbidium* are fusiform. The capsules of *Acanthephippium* and *Geodorum* are relatively broad, oblong or ellipsoid, 4–5 cm. long, 2–2.5 cm. in diameter. Variations in the shapes and sizes of the capsules depend primarily upon the proportional increases in length and in diameter of the young fruits as they develop from the ovaries to mature capsules, and secondarily upon the sizes of the ovaries. These conditions are illustrated by the following specific data:

Species	Ovaries in mm.	Capsules in mm.	Times of increase in	
			Length	Diameter
Acampe multiflora	5 × 2	70 × 5	14	3–4
Geodorum densiflorum	6 × 1	40 × 20	6–7	20
Cleisostoma teres	8 × 1	40 × 6	5	6
Spiranthes hongkongensis	3 × 2	6 × 2.5	2	2–3
Goodyera procera	3 × 1	5 × 2	1–2	2

It is obvious that in the development of the unusually long capsule of *Acampe* there is proportionally a large increase in the length of the fruit. In contrast, in the broadly oblong or ellipsoid capsule of *Geodorum*, there is a tremendous increase in diameter. In the fusiform capsule of *Cleisostoma* there is a moderate increase both in length and in diameter. In the development of small capsules in *Spiranthes* and *Goodyera* there is little increase in size from an ovary to a capsule. Actually in these cases, the lower capsules of a spike or raceme are mature while the upper flowers of the same inflorescence are opening. The ovaries of several species of *Habenaria* in Hong Kong are narrowed at the apex. Consequently the capsules of the species are ellipsoid and strongly pointed at both ends.

Most of the capsules observed are erect. The capsules of *Calanthe masuca, Geodorum,* and *Phaius* are pendulous. The capsule of *Arundina* is at a right angle to the rachis.

The persistent columns of *Arundina, Coelogyne, Cymbidium, Geodorum,* and *Phaius* become stiff and beak-like in fruit. These capsules are strongly rostrate. The apices of the small capsules of *Appendicula, Goodyera, Malaxis, Spiranthes,* and *Zeuxine* are crowned with the remains of the partial or complete perianth segments. The capsules of all 23 genera of Hong Kong orchids observed are appendaged either by the persistent column alone or with some remains of the perianth.

SEED AND SEEDLING

Orchids are well known to be big producers of small seeds. Over one hundred years ago, Charles Darwin reported that in the capsule of a British orchid, *Coeloglossum*, there is an average of 6 200 seeds. In 1956, L. Knudson reported that a capsule of *Cattleya* produced an average of 256 000 seeds. Other orchidologists have higher counts with other genera. All orchids produce numerous dust-like winged seeds. These are adapted for wind dissemination. They contain neither a differentiated embryo, nor food reserved for the development of the content.

An orchid seed is minute, fusiform, with a dark central spot of densely protoplasmic cells and a reticulate seed coat of one cell layer thick which extends at both ends into tiny wings. In nature an orchid seed does not germinate unless it is first invaded by a beneficial symbiotic fungus which obtains its own food through decomposition of the surrounding material and provides nutrition for the initial growth of the young seedling. Modern technology has enabled orchid hybridizers to raise orchids from seed in culture-flasks by providing suitable culture media. It takes a period of five or six years to raise an orchid from seed to flower.

II
Keys and Descriptions

The aim of this chapter is to provide a set of keys to enable the reader to identify the genera of orchids in Hong Kong. With the generic descriptions and the keys to the species in each genus, the reader should be able to determine the species common in this area.

In accordance with the existing *International Code of Botanical Nomenclature* the names for the subfamilies and tribes used here are different from those found in older references in orchidology that follow Schlechter's system of classification and terminology. For the convenience of cross reference, the familiar Schlecterian names are given in parentheses in the keys.

KEYS TO THE SUBFAMILIES, TRIBES, AND GENERA

Key to the Subfamilies and Tribes

A. Lip slipper-like; lateral sepals completely connate, forming a foliaceous synsepalum hidden by the lip; column terminated by a shield-like or a butterfly-like staminode; anthers 2, on the sides of the column; pollen grains dusty, not consolidated into pollinium......
....**Subfam. I. CYPRIPEDIOIDEAE** (Diandrae)
Tribe A. Cypripedieae (Cypripediloideae)
(Single genus) 1. *Paphiopedilum*

AA. Lip various, never slipper-like; lateral sepals more or less separated, usually spreading sidewise, not hidden by the lip; column terminated by a functional anther; anther 1, in front or at the apex of the column; pollen grains agglutinated into 2–8 pollinia......
....**Subfam. II. ORCHIDOIDEAE** (Monandrae)

   B. Plants with solitary herbaceous erect and deciduous stem, and fleshy tubers or corms, without rhizomes or pseudobulbs; column short and stout, bearing two widely separated thecae on the front side; anther closely adnate to the column by the broad bases of the thecae; pollinia sectile...............**Tribe B. Orchideae** (Ophrydoideae)

   BB. Plants rhizomatous, pseudobulbous, or with erect stems bearing many distichously arranged leaves, rarely aphyllous (in *Aphyllorchis*); columns various, each bearing a terminal anther; anther attached to the column by a short and thin filament; pollinia various.

      C. Stems modified into creeping rhizomes with fleshy elongated internodes, or into short erect rootstocks covered by the bases of the petioles, without pseudobulb or continuous apical growth, rarely appearing parasitic or saprophytic and without leaves (*Aphyllorchis*); pollinia granular or sectile; anther cap persistent...
....**Tribe C. Neottieae** (Polychondreae)

19

cc. Stems various, either modified into slender stringy rhizomes each terminated by a pseudobulb, or into ovoid, conic, cylindric, petiole-like, or irregularly compressed lumpy pseudobulb, or erect with numerous distichously arranged leaves; pollinia waxy or cartilagious; anther cap deciduous...
....**Tribe D. Epidendreae** (Kerosphaereae)

Key to the Genera of Tribe B. Orchideae

A. Plants 15 cm. or more tall, with fleshy subterranean tubers; lateral sepals free, neither connate at the base nor saccate at the middle; rostellum yoke-shaped, collar-like, beak-like, or folded, not stipitate; lip undivided or trilobed, not folded at the middle.

  B. Flowers large, showy, 5–6 cm. across; lateral sepals spreading upward, the inner margin connivent with the back of the odd sepal; lateral lobes of the lip fringed; spurs tubular, 10 cm. or more long...2. *Pecteilis*

  BB. Flowers small and inconspicuous, or rarely medium-sized, 0.5–2.5 cm. across; lateral sepals spreading sidewise, not touching the back of the odd sepal; spurs globular, oblong, or cylindric, 0.2–4 cm. long.

    c. Lips undivided; stigma 1.

      D. Bracts foliaceous; lip pink, obcordate; petals and odd sepal connivent, forming a galea over the column; rostellum folded, beak-like............3. *Brachycorythis*

      DD. Bracts membranous, small, lanceolate; lip yellowish green, tongue-like; petals and sepals separated; rostellum collar-like..................4. *Platanthera*

    cc. Lips trilobed; stigmas 2.

      D. Flowers red, white, bright yellow, or greenish yellow; lateral sepals spreading sidewise; back of column upright; thecae divergent; rostellum large, projecting upward from the back of the column, yoke-like or beak-like, with long arms; stigmas below the rostellum-arms; caudicles slender, equal or much longer than the pollinia..5. *Habenaria*

      DD. Flowers green or white and tinted green; lateral sepals suberect, close to the column; back of column slanting; thecae parallel and close; rostellum small, folded, with short arms; stigmas in front of the rostellum-arms, adnate to the base of lip; caudicles shorter than the pollinia..........................6. *Peristylus*

AA. Plants small, tender, less than 10 cm. tall, with subglobose corms; lateral sepals connate at the lower portion, saccate at the middle; rostellum stipitate; lip very narrow, folded at the middle and extending forward into a pendulous slender front lobe 1 mm. in diameter...7. *Disperis*

Key to the Genera of Tribe C. Neottieae

A. Plants normal, with phytosynthetic green leaves; flowers various; rachis, pedicels and ovary not violet purple.

  B. Leaves solitary, suborbicular and cordate at base, appearing subsessile and lying flat on the ground; plant bearing slender underground stolon terminated by a subglobose corm; flowers 2, in a terminal umbelliform cluster; pollinia without caudicle and viscidium..8. *Nervilia*

  BB. Leaves 2 or more, petiolate, linear, lanceolate, elliptic or ovate, the base cuneate, acute, obtuse or rarely oblique subcordate; plants with short erect rootstocks or creeping rhizomes; flowers several to many, in a spike or raceme; pollinia with viscidium normally.

    c. Plants acaulescent; rootstocks short and erect, concealed by the base of the petioles; pollinia finely granular, without obvious caudicles.

      D. Flowers non-resupinate; sepals, petals, and lip spreading; lip uppermost in position, flat, ovate, acute; stigma large, covering the entire front of the column..9. *Cryptostylis*

      DD. Flowers resupinate; sepals, petals, and lip overlapping; lip lowermost in position, concave and concealed by the sepals and petals; stigma (or stigmas) limited to small area of the column.

E. Leaves linear, appearing with the flowers; stigma 1, with a defined margin, suborbicular or shield-like and tricuspidate at the upper margin; viscidium concealed below the rostellum between the lobes, or obscure; lip with 2 spherical calli, one on each side of the base.......10. *Spiranthes*

EE. Leaves elliptic, appearing after anthesis; stigmas 2, widely separated on a plane surface of the column, oblong; viscidia exposed at the apex of the lobes of the rostellum; lip unguiculate, with 2 subulate and hooked appendages, one on each side of the basal claw..............11. *Manniella*

CC. Plants with erect leafy shoots; rhizomes creeping, with slender internodes; leaves with sheathing petioles; pollinia sectile, with slender caudicles.

D. Stems herbaceous; rhizomes succulent or fleshy; leaves flat, smooth, without elevated nerves beneath; inflorescence terminal to a leafy shoot.

E. Rhizomes succulent, worm-like, 4–6 cm. long, clinging on wet boulders by dense hairs; sepals connate above the middle forming a tube..........
....12. *Cheirostylis*

EE. Rhizomes fleshy, slender, 10 cm. or more long, creeping over wet forest floor by rather stout roots; sepals distinct, the lateral ones spreading.

F. Flowers with the lip fringed and exposed in a horizontal position; spur conic, cleft at the apex; odd sepals and petals strongly recurved.
....13. *Anoectochilus*

FF. Flowers with small lip concealed by the sepals and petals, if exposed then twisted and not fringed; odd sepals and petals ovate and concave, not reflexed at the apex.

G. Lip twisted, with slender claw; column slender, naked, twisted, without lateral glands; stigma 1, on plane surface in front of the column..14. *Ludisia*

GG. Lip ventricose, saccate or calcarate, not twisted, without a claw; column stout, with or without lateral glands; stigma 2, or 1 and below the rostellum.

H. Flowers resupinate, the odd sepal uppermost; lip ventricose, the apical portion exposed, undivided and reflexed, or divergently bilobed.

I. Lip entire, the apex reflexed, the base ventricose and papillose on the inside; stigma 1; column without lateral glands..15. *Goodyera*

II. Lip constricted at the middle, the apical portion divergently bilobed or obovate, the base saccate and with inflexed margins, the sac containing 2 calli; stigmas 2; column with lateral glands........................16. *Zeuxine*

HH. Flowers non-resupinate, the odd sepal lowermost; lip concealed by the sepals and petals, concave, the apex obtuse or acute.

I. Flowering bract shorter than the ovary; lip ventricose or saccate, the sac with erect papillae or stipitate calli; column appendaged in front; stigmas connivent below the rostellum..17. *Hetaeria*

II. Flowering bract longer than the ovary; lip calcarate, the spur oblong, pointing backward, with 2 hanging calli pointing to the tip; column short, stout, without front appendages; stigmas 2, not meeting below the rostellum.
....18. *Vrydagzynea*

DD. Stems tough, bamboo-like; rhizomes woody; leaves plicate, with numerous elevated nerves beneath; inflorescences axillary to leaves..........19. *Tropidia*

AA. Plants appearing parasitic, aphyllous, the stem purple-streaked; flowers small, yellow, hardly open; rachis, pedicels, and ovary violet-purple..........................20. *Aphyllorchis*

Key to the Genera of Tribe D. Epidendreae

A. Plants with sympodial growth; terrestrial or epiphytic orchids with the roots emerging from subterranean rootstock or aerial rhizome (except in *Epidendrum*, from leafy stem); pollinia 4 or 8, rarely 2 or 6, with various devices of attachment (for AA see page 25).

  B. Stems slender, erect or hanging, caespitose; leaves many, distichous or loosely arranged on the stem.

    C. Plants bamboo-like, with 4–8 large plicate-nervose chartaceous leaves 15–40 cm. long, 4–7 cm. wide; flowers yellow, medium-sized, 30–40 in a raceme on elongated scapes up to 40 cm. long...21. *Cephalantheropsis*

    CC. Plants with various habit, with leaves coriaceous, smooth, less then 12 cm. long and 3 cm. wide; flowers red, white, purplish-pink, rarely yellow in some cultivated forms, 15 or less in a raceme on short scapes.

      D. Racemes emerging laterally on mature stems, occasionally after the leaves dropped off; lateral sepals adnate to the column-foot forming a mentum.

        E. Flowers small, hardly opening; lip fleshy, with a pendulous appendage at the base and an unicorn-like callus on the disc; pollinia 6 in 2 groups of 3, attached to slender caudicles.............................22. *Appendicula*

        EE. Flowers large, with spreading sepals and petals; lip without a basal appendage, the disc lamellate or ridged; pollinia 4, without caudicle or obvious viscidium..23. *Dendrobium*

      DD. Racemes terminal to a leafy stem; sepals free, not adnate to the column foot to form a mentum.

        E. Spontaneous terrestrial orchids with all roots attached to the base of the stem and hidden in the soil; leaves lanceolate; flowers opening one at a time; lip embracing the column but free from it; pollinia 8, the connecting appendage and viscidium obsolete...................24. *Arundina*

        EE. Cultivated epiphytic orchids with elongated aerial roots attached to the supporting object; leaves oblong; several flowers opening simultaneously; lip adnate to the sides of the column, forming a pouch below the column; pollinia 4, with well developed granular appendage associated with a viscidium against the rostellum................................25. *Epidendrum*

  BB. Stems stout or thickened, specialized into tuberous or conic rhizomes or to slender rhizomes each with one joint enlarged into a pseudobulb; leaves 1–4, terminal to the annual growth, usually to a pseudobulb.

    C. Rhizomes stout and short, the internodes shorter than the thickness of the same, or rhizomes fleshy with elongated succulent internodes; leaves produced 1 a year, terminal to the annual growth of the stem, ovate, cordate or linear; anthers obovoid, bicornute (except *Ania*).

      D. Leaves ovate or cordate; pseudobulbs petiole-like, slender, elongated.

        E. Rhizomes fleshy, bearing few roots; leaves ovate, varietated; flowers non-resupinate, without a mentum; lip calcarate, the spur enlarged at the end..26. *Nephelaphyllum*

        EE. Rhizomes cord-like, bearing numerous roots; leaves cordate, not variegated; flowers resupinate, with a mentum; lip not spurred.................
....27. *Mischobulbum*

      DD. Leaves linear or elliptic-linear; pseudobulbs ovoid or oblong-conic.

        E. Pseudobulbs cylindric-conic; flowers with short mentum; lip not spurred; pollinia discoid..28. *Tainia*

        EE. Pseudobulbs ovoid; flowers without mentum; lip spurred; pollinia unequal, obovoid...29. *Ania*

    CC. Rhizomes either wire-like with slender internodes and terminated by an ovoid, oblong or ellipsoid pseudobulb, or conic, cylindric-conic, or tuberous with concentric rings; leaves 2 to several, terminal to the pseudobulb; anthers obovoid or hemispherical, not bicornute.

      D. Plants growing on exposed grassy slopes, passing the dry winter season in a leafless dormant state buried in the soil; rhizomes subterranean, tuberous, with concentric rings.

E. Flowers appearing before the leaves; lip saccate, calcarate, or associated with a short mentum.

    F. Rhizomes cylindric or yoke-shaped, 1–1.5 cm. in diameter, buds apical, developing either to a flowering or a leafy shoot; flowers with a short mentum; sepals and petals hairy; pollinia 8, obovoid and acuminate..30. *Pachystoma*

    FF. Rhizomes obovoid or oblong, 3–8 cm. in diameter, the buds of the flowering and the vegetative shoots close or separated irregularly; sepals and petals not hairy; pollinia 2, subglobose, on a short stipe..31. *Eulophia*

EE. Flowers appearing with the new or the well developed leaves; lip without spur and not associated with mentum.

    F. Leaves basal to a flowering shoot; scapes glabrous; flowers purple-pink (white in a cultivar); midlobe of lip lamellate...............32. *Bletilla*

    FF. Leafy and flowering shoots separated; scapes pubescent; flowers bright yellow; midlobe of lip with 3 short keels........33. *Spathoglottis*

DD. Plants either epiphytic on exposed boulders or growing in shady wet or damp habitat along streams under trees, normally leafy throughout the year; rhizome conic, cylindric-conic, or wire-like and terminated by a pseudobulb.

    E. Flowers hardly open, the sepals and petals connivent; lip 2 mm. long, undivided, concave; pollinia 8 in two groups, on an elongated stipe with a viscidium...34. *Thelasis*

    EE. Flowers widely open, the sepals and petals spreading; lips unlike the above; pollinia 2 on very short stipe, 4 without stipe, or 8 without stipe.

        F. Flowers small and numerous, 30 or more in a raceme terminal to a leafy stem changing into a conic or cylindric-conic, often many jointed pseudobulb; lip either spathulate, recurved and cuneate at the base, or flat, concave and auriculate at the base.

            G. Lip spathulate, recurved from the middle, cuneate and with 2 calli at the base; column slender, arcuate and winged at the apical portion...35. *Liparis*

            GG. Lip flat and auriculate, or the sides erect and the base auriculate; column very short, partially concealed in the depression at the base of the lip..36. *Malaxis*

        FF. Flowers medium-sized or large and showy (except in some species of *Bulbophyllum*), one or more (up to 20) in a raceme; lip neither spathulate and cuneate, nor flat, concave and auriculate at the base.

            G. Plants creeping; rhizomes slender, usually terminated by a pseudobulb; new shoot emerging below the pseudobulb.

                H. Flowers large and showy; lip trumpet-shaped, the margin wavy or crisped; leaves linear, band-like, 20–30 cm. long, pseudobulb cylindric, ellipsoid, 10 cm. or more long.........
....37. *Cattleya*

                HH. Flowers medium-sized or small; lip not trumpet-shaped; leaves not band-like, usually oblong, elliptic, rarely linear; pseudobulb ovoid or oblong, less than 10 cm. long.

                    I. Leaves 2 to each pseudobulb; scape situated between the leaves and apical to the pseudobulb; flowers without mentum.

                        J. Scape without basal scales; pseudobulb ovoid; flowers usually 15 or more in a hanging or erect spike, appearing with the young shoot before the formation of the pseudobulb and the expansion of the leaves.......................................38. *Pholidota*

                        JJ. Scape with basal scales; pseudobulb oblong; flowers few, usually 2 or 3 on an erect scape, appearing between two coriaceous mature leaves terminal to a pseudobulb.....................................39. *Coelogyne*

ii. Leaf 1 or leaves several to each pseudobulb; scape emerging from the base or the side of a pseudobulb; flowers with a mentum (except *Sophronitis*).

    J. Pollinia 4 in 2 pairs; lip fleshy, tongue-like, without sidelobes, very mobile; leaf 1 to each pseudobulb.

        K. All flowers facing one direction in a fan-shaped umbel; lateral sepals 2–6 times as long as the odd sepal, twisted, with the upper margins meeting and coherent to form a platform before the lip................................40. *Cirrhopetalum*

        KK. Flowers facing all directions in a subumbelliform raceme; sepals almost similar in size and shape, the upper margins of the lateral sepals not coherent beyond the mentum....41. *Bulbophyllum*

    JJ. Pollinia 8 in 2 groups of 4; lip 3-lobed, the sidelobes erect, not very mobile; leaves several (rarely 1) to each pseudobulb.

        K. Flowers greenish white or creamy white, occasionally tinged pink, with a mentum; midlobe of lip expanded, rotundate or 2-lobed. ....42. *Eria*

        KK. Flowers cardinal red; lateral lobes of the lip overlapping over the column; midlobe of the lip linguiform; no mentum...............43. *Sophronitis*

GG. Plants caespitose, without slender creeping rhizome; new shoot emerging from the base of a conic or cylindric-conic pseudobulb with concentric rings.

    H. Flower with a mentum; sepals coherent, forming an oblique ventricose-urceolate tube enclosing the petals and lip... ....44. *Acanthephippium*

HH. Flowers without mentum; sepals and petals spreading.

    I. Lip spurred; pollinia 8, obovoid, with or without caudicles, not on a stipe; leaves chartaceous, plicate-venose.

        J. Spur conic, shorter than the ovary; lip attached to the base of the colum, trumpet-shaped, the sides overlapping over the column and anther, the margin wavy...45. *Phaius*

        JJ. Spur slender, longer than the ovary; lip 3-lobed, adnate to the sides of the column forming a pouch. ....46. *Calanthe*

    II. Lip ventricose, saccate or contracted at the base, 3-lobed and not spurred; pollinia 2, deeply cleft and often separated into 4, on a short stipe; leaves coriaceous, or chartaceous and plicate-venose.

        J. Pseudobulb strongly compressed ovoid; flowers solitary, the sepals 3 cm. wide...............47. *Lycaste*

        JJ. Pseudobulb subglobose or ovoid-conic; flowers several to many in a raceme, the sepals less than 2 cm. wide.

            K. Plants with subglobose, ellipsoid or conic pseudobulbs; midlobe of lip recurved or lip undivided, concave, without an obvious midlobe.

                L. Scape bending at anthesis; raceme nutant; flowers crowded; lip ventricose, with calli in the sac; stipe of pollinia slender............ ....48. *Geodorum*

ᴌᴌ. Scape straight at anthesis; raceme erect; flowers loosely spaced; lip not ventricose, keeled and without calli; stipe of pollinia broad and short...................49. *Cymbidium*

ᴋᴋ. Plants with suberect stems bearing equitant sheaths or leaves, and terminated with one or two laterally compressed pseudobulbs each with 1 or 2 leaves at the top; midlobe of the lip greatly widened, suborbicular, notched or lobed at the apex.

ʟ. Flowers numerous, in a long raceme or panicle; lip unguiculate, narrowed at the base; column petaloid.

ᴍ. Lip straight; disc cristate; flowers paniculate......................50. *Oncidium*

ᴍᴍ. Lip bent with the narrow base parallel to the column and the enlarged portion pointing downward; disc with 2 calli; flowers racemose........51. *Odontoglossum*

ʟʟ. Flowers 3–5 in a loose raceme; lip flat, auriculate at the base; disc slightly lamellate; column not petaloid...............52. *Miltonia*

ᴀᴀ. Plants with monopodial growth; epiphytic orchids with aerial roots emerging from the leafy stem; pollinia 2 or 4, on elongated or specially modified stipe (for ᴀ see page 22).

ʙ. Flowers ephemeral; sepals and petals lanceolate, acute at the apex; pollinia strongly unequal, oblong, on short and stout stipe with very narrow viscidium..................
....53. *Thrixspermum*

ʙʙ. Flowers lasting for several to many days; sepals and petals oblong, ovate or rotundate, if linear then with rounded tips; pollinia appearing subspherical, on elongated stipe.

ᴄ. Flowers red, or red and yellow; sepals and petals narrow, 4 or 5 times longer than wide, enlarged at the apices, maculated.

ᴅ. Lip movable; disc without tuberculate calli between the lateral lobes.........
....54. *Arachnis*

ᴅᴅ. Lip not movable; disc with 2 tuberculate calli between the lateral lobes...
....55. *Renanthera*

ᴄᴄ. Flowers of various color but not red; sepals and petals as wide as long, if narrower then less than three times longer than wide.

ᴅ. Lip without spur or sac; midlobe bifid at the apex, the segments usually filiform.

ᴇ. Racemes shorter than the leaves; flowers small, yellow, less than 2 cm. across; lip with flat smooth disc; leaves 5–8 cm. long, well spaced on the stem (native)..56. *Diploprora*

ᴇᴇ. Racemes or panicles longer than the leaves; flowers white or purple-pink, large, showy, 4–6 cm. across; lip with tuberculate callus between the lateral lobes; leaves various (cultivated)......................57. *Phalaenopsis*

ᴅᴅ. Lip spurred or saccate; midlobe rotundate, acute, obtuse, without filiform segments, or bilobed and fringed in *Ornithochilus*.

ᴇ. Plant aphyllous; flowers small, inconspicuous, 2–3 mm. long.............
....58. *Microcoelia*

ᴇᴇ. Plants with green phytosynthetic leaves; flowers 1–8 cm. across.

ꜰ. Sepals and petals maculate, with horizontal purplish bars; lip velvety-papillose, shallowly saccate at the base; leaves large, fleshy, 2.5–4.5 cm. wide..59. *Acampe*

ꜰꜰ. Sepals and petals without horizontal bars; lip ventricose, conic-saccate, or prominently calcarate; leaves of moderate size (except *Vanda tricolor*), or terete, 0.4–2.5 cm. wide.

ɢ. Flowers large, showy, 5–8 cm. across; petals suborbicular narrowed at the base, or linear-oblong and unguiculate; plants 1–2 m. high (except *V. tricolor*)..............................60. *Vanda*

GG. Flowers rather small, 1–2 cm. across; petals oblong, not unguiculate; plants of moderate size, 10–40 cm. high.

H. Flowers 2–3 in a small cluster; lip ventricose, without callus or keel; pollinia 2, appearing spherical from the front, concave at the back...............................61. *Gastrochilus*

HH. Flowers many in a raceme or numerous in a panicle; spurs of lip conic, calcarate or subspherical, with callus on the back of the throat and a keel running down from the disc to the apex of the spur (except *Ornithochilus*); pollinia 2 or 4.

I. Midlobe of the lip neither lobed nor fringed; pollinia 4; orifice of spur with a callus on the back wall.

J. Leaves linear-oblong, erose at the apex; spur enlarged and spherical at the apex; anther acuminate and beaked at the apex; column with 2 arms at the apex..62. *Robiquetia*

JJ. Leaves terete, lanceolate or band-like, the apex acute, obtuse or 2-lobed, not erose; spur obtuse, not enlarged at the tip; anthers various; column with 2 teeth at the apex.....................63. *Cleisostoma* (Sarcanthus)

II. Midlobe of the lip prominently bilobed and fringed; pollinia 2; orifice of spur with a flap on the front wall..64. *Ornithochilus*

Descriptions of Genera with Illustrations and Keys to the Species

1. **Paphiopedilum** Pfitzer 兜蘭屬 (Figure 5)

Flowers rather large and showy, solitary or in a loose raceme; odd sepal uppermost, large, colorful, lateral sepals connate forming a synsepalum generally hidden behind the lip; petals narrower than the odd sepal, spreading out horizontally; lip slipper-shaped, the margin of the basal half incurved, the base of the pouch raised slightly into an auricle on each side; column short, terete, terminated by a foliaceous staminode; fertile anthers 2, on the sides of the column, sessile, subglobose, locules 2, parallel; pollen grains dusty and in a sticky mass, not consolidated into pollinium; stigma discoid, on a short stalk; ovary cylindric. Capsules fusiform. Terrestrial or epiphytic orchids with sympodial growth. New shoots arising from the base of an old one. Leaves oblong-linear, or band-like, distichous, uniformly green or tessellated with darker and paler green. Scape terminal to a leafy shoot. One species native to Hong Kong, and a few others introduced from tropical Asia for ornamental purposes.

Key to the Species

A. Scapes 40 cm. or less tall; flowers solitary.

B. Leaves linear-oblong, 6–10 cm. long; synsepalum shorter than the lip; scape, ovary, and the bracts obviously pubescent.

C. Pedicel and ovary together shorter than the lip; petals ciliate; staminode deeply lobed at the apex; margin of the slipper smooth (native of Hong Kong).........
....*P. purpuratum* Pfitzer

CC. Pedicel and ovary together longer than the lip; petals not ciliate; staminode shallowly notched at the apex; upper margin of slipper with a tooth on each side (native of Thailand)..*P. appletonianum* Rolfe

BB. Leaves band-like, 20–30 cm. long; synsepalum equal or longer than the lip; scape, bract, and ovary not obviously pubescent (native of Thailand)............*P. exul* Rolfe
AA. Scape 60 cm. or more tall; rachis branched; flowers 3–5 in a loose raceme (native of the Philippines) .... *P. haynaldianum* (Reichb. f.) Pfitzer

FIGURE 5. *Paphiopedilum purpuratum*: A. The habit sketch of a flowering plant showing variegated leaves and a scape with 1 flower subtended by a large bract. B. The lateral view of a flower with one-half of the dorsal sepal, 1 petal and one-half of the lip removed, showing the butterfly-like staminode apical to the column, and the inferior hairy ovary. C. A portion of the column showing the staminode, one functional anther with pollen mass, and the fleshy style terminated by a discoid stigma. D. A young capsule.

2. **Pecteilis** Rafin. 白蝶花屬 (FIGURE 6)

Flowers large, showy, in a terminal raceme; sepals ovate, spreading, all pointing upward, the lateral one connivent with the odd sepal; petals small, linear; lip 3-lobed, calcarate, the spur twice or more times longer than the ovary and pedicels together,

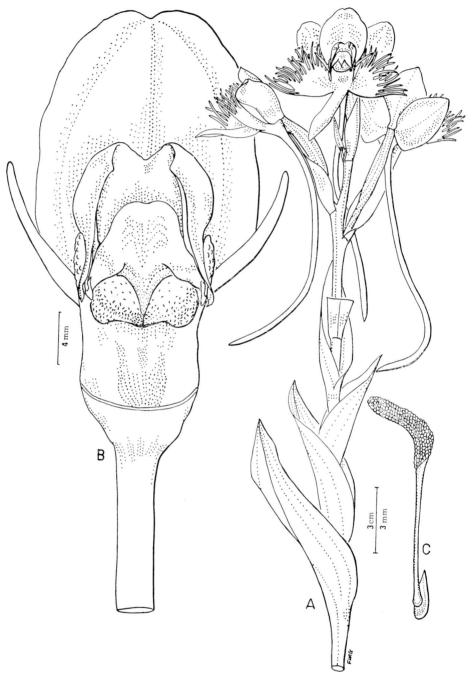

FIGURE 6. *Pecteilis susannae*: A. The habit sketch of the flowering shoot showing the cauline leaves, the front and lateral views of the flowers, the sepals spreading upward and connivent, the trilobed lip with fringed lateral lobes and narrow midlobe and the prolonged spurs. B. The front view of a flower with the lip and the lateral sepals removed, showing the dorsal sepal, the 2 horn-like petals, the column with a widely separated thecae, the concealed viscidia, the lateral glands, the fleshy large rostellum, the stigmas, and the orifice of the spur. C. A pollinium on slender caudicle and a viscidium.

the lateral lobes spreading, fringed, the midlobe linear, entire, obtuse; column broad, shield-like, retuse at the apex; roestellum fleshy, flat, broadly triangular, upright, rounded and notched at the apex, with long arms; anther cells divergent; pollinia 2, finely tessellate; caudicles slender, elongated; viscidia concealed; stigmas 2, below the rostellum, free from the auricles, grooved down the middle. Terrestrial orchid, requiring dormancy in the cool and dry winter season. Leaves cauline, ovate-elliptic, up to 11 cm. long, 4.5 cm. wide, entire. One native species, *P. susannae* (L.) Rafin., occurring on the grassy slopes of high mountains; also widespread in southeastern Asia.

3. **Brachycorythis** Lindl. 短盔蘭屬 (FIGURE 7)

Flowers medium-sized, subtended by foliaceous bracts or solitary in the axils of normal leaves; sepals free, odd one uppermost, ovate, concave, the lateral sepals spreading, oblique; petals small, below the odd sepal and with it forming a hood over the column; lip undivided, fleshy, obcordate, calcarate, the spur stout, short, bilobed at the apex; column short and stout, lateral glands prominent; rostellum yoke-shaped; anther 2 celled, the thecae proximate, parallel; pollinia 2, coarsely tessellate; caudicles short; viscidia large, exposed; stigma one, large, occupying the front cavity of the column below the rostellum, glossy; ovary sessile, striate-sulcate, the ridges papillose. Capsules erect, narrowly oblong. Terrestrial orchids requiring dormancy in the cool dry season, with fleshy oblong tubers. Stem leafy. Leaves lanceolate, 3–7 cm. long, sessile. One species growing among short grasses on windy peaks of high mountains in Hong Kong, *B. galeandra* (Reichb. f.) Summerh.

4. **Platanthera** L. C. Rich. 長距蘭屬 (FIGURE 8)

Flowers small, in a terminal spike; sepals free, the odd one uppermost, ovate, concave, the lateral sepals reflexed; petals oblique ovate, the base of the lower side subcordate, apex acute or obtuse; lip undivided, tongue-like, hanging, calcarate; column short and stout, deeply retuse at the apex, the glands prominent; rostellum horizontal, collar-like, fleshy; anther 2-celled, the thecae divergent, widely separated; pollinia 2, finely tessellate, on slender caudicles; viscidia discoid; stigma situated above the lip below the rostellum, discoid, deeply depressed in the middle; ovary bending and slightly narrowed at the apex, the ridges winged. Capsules ellipsoid. Terrestrial orchids, requiring dormancy in the winter dry and cool season, with fusiform tubers each bearing a subterminal bud. Stem solitary, leafy. Leaves one at the ground level and 1 or 2 cauline, bract-like. Three species growing on windy grassy slopes or ravines of mountains in Hong Kong.

Key to the Species

A. Spur curved downward with the apex pointing to the ground; petals equal or shorter than the odd sepal and with their upper margins connivent with it.
  B. Lower leaves linear-lanceolate, 8–10 cm. long, 1.3–1.5 cm. wide; spur 13–14 mm. long, parallel and then curved away from the ovary; ridges of the ovary hardly winged; lip 9 mm. long, 3–3.5 mm. wide......................*P. minor* (Miq.) Reichb. f.
  BB. Lower leaves oblong or rarely ovate-oblong, 9–15 cm. long, 2–3.7 cm. wide; spur 8–12 mm. long, parallel and close to the ovary, often with the tip projecting in the axil of the pedical; ridges of the ovary prominently winged; lip 6 mm. long, 2.5–3 mm. wide..*P. angustata* (Bl.) Lindl.
AA. Spur curved upward with the apex pointing to the sky; petals longer than the odd sepal and with their apices pointing away from it........................*P. mandarinorum* Reichb. f.

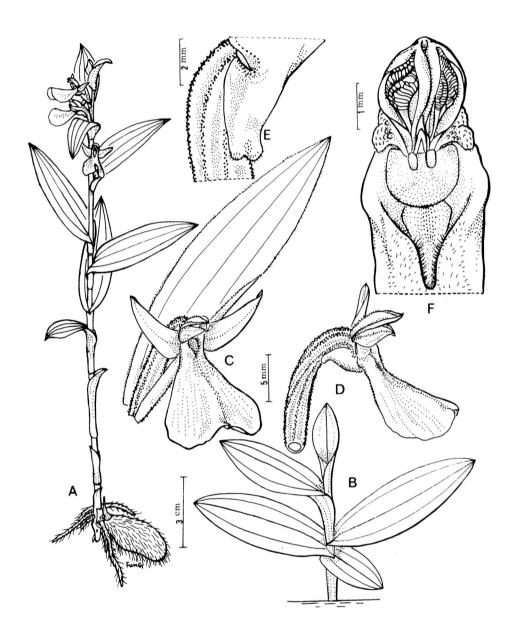

FIGURE 7. *Brachycorythis galeandra*: A. The habit sketch of a flowering plant showing subterranean tuber, the leafy stem, the position of the flowers axillary to normal leaves. B. The
habit sketch of a young plant before anthesis showing a convolute young leaf. C. The front
view of a flower and the subtending leaf showing the widespread sepals, the petals and the
undivided lip. D. The lateral view of a flower with 1 sepal removed, showing the dorsal sepal,
the petals and the saccate spur of the lip. E. The front view of a spur showing the 2 lobes at
the apex. F. The front view of an apical portion of the column showing the parallel thecae,
the pollinia with caudicles and exposed viscidia, the bent rostellum, the lateral glands, a
large stigma and the orifice of the spur.

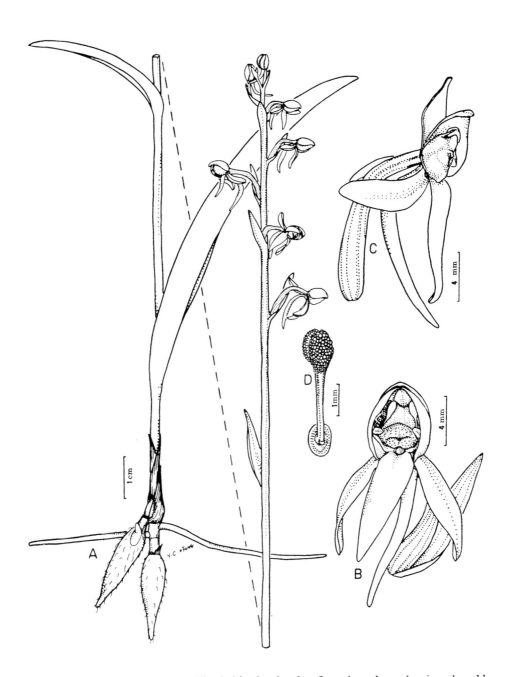

FIGURE 8. *Platanthera minor*: A. The habit sketch of a flowering plant showing the old tuber from which the flowering shoot was developed, the position of the new tuber with a bud for the next year's growth, the cataphylls, the long narrow leaves and the flowering scape (detached). B. The front view of a flower showing the relative positions of the sepals, the petals, the undivided tongue-like lip, the spur, the column with 2 divergent thecae, the horizontal fleshy rostellum, the orifice of the spur, and the stigma. C. The lateral view of a flower with one-half of the dorsal sepal and a petal removed, showing the lateral sepals curving backward, the tongue-like lip curving downward, the long spur, and the very broad column with a lateral gland. D. A pollinium on a slender caudicle attached to a discoid viscidium.

5. **Habenaria** Willd. 玉鳳花屬 (Figure 9)

Flowers medium-sized or small, in terminal racemes; sepals subequal, free, the odd one ovate, concave, the lateral sepals spreading sidewise or twisted and curved below the lip; petals usually smaller than the sepals; lip trilobed, calcarate, spur longer than the ovary, the lateral lobes usually wider than the midlobe, spreading; column short, the back erect, the base broad and with a glandular auricle on each side; rostellum large, fleshy, yoke-shaped or beak-like, with slender long arms; anther 2-celled, thecae widely separated, broadly attached to the front of the column; pollinia 2, sectile, each on a slender caudicle with viscidium; stigmas 2, slightly curved below the rostellum-arms; ovary clyindric, slightly narrowed at the apex, striate-sulcate. Capsules fusiform or ellipsoid, pointed at both ends. Terrestrial orchids requiring dormancy during the dry cool season; with fleshy subterranean tubers; growing in shade along streams, or among grasses and sedges of exposed slopes of high mountains. Stems leafy, herbaceous, deciduous. Leaves convolute, tender, elliptic or lanceolate-elliptic, rarely ovate. Inflorescences racemose, terminal to leafy shoots. Six species in Hong Kong.

Key to the Species

A. Flowers medium-sized, showy, red, bright yellow, or white, 1.5–3 cm. across; spurs 3–4.5 cm. long.

   B. Flowers red; lateral sepals twisted at the base and curved behind the lip; rostellum beak-like; growing in shade along stream...............................*H. rhodocheila* Hance
   BB. Flowers white or yellow; lateral sepals spreading sidewise; rostellum fleshy, yoke-shaped.

   　　C. Lateral lobes of the lip wider than the midlobe, irregularly dentate; flowers white; leaves elliptic, up to 4 cm. wide..................*H. dentata* (Sw.) Schlechter
   　　CC. Lateral lobes of lip narrower than the midlobe, entire; flowers bright yellow; leaves linear-lanceolate, 1.5 cm. or less wide................................*H. linguella* Lindl.

AA. Flowers small, green or yellowish green, 1 cm. or less across; spurs 2 cm. or less long.

   B. Leaves broad-ovate, subcordate, 2–3.5 cm. long and wide; flowers 8 mm. across; spur shorter than the ovary...*H. reniformis* (D. Don) HK. f.
   BB. Leaves linear-lanceolate, or ovate-lanceolate, 5–13 cm. long, 1–2 cm. wide; flowers 10 mm. across; spur longer than the ovary.

   　　C. Bracts and rachis glabrous; leaves linear-lanceolate, 1–1.5 cm. wide; spur 1 cm. long; rachis 7–10 cm. long...*H. leptoloba* Benth.
   　　CC. Bracts and rachis pilose; leaves ovate-lanceolate, 3 cm. wide; spur 2 cm. long, or longer; racemes 30 cm. long.......................................*H. ciliolaris* Kränzlin

---

Figure 9. *Habenaria dentata*:   A. The habit sketch of a flowering plant showing the subterranean tubers, the fibrous roots, the cauline leaves and the flowering scape.   B. A flower enlarged showing the relative positions of the sepals, the petals, the trilobed lip with long, large, irregularly dentate lateral lobes, linear midlobe and cylindrical long spur, the column with 2 thecae, the lateral glands, the viscidia on the rostellum-arms, the prolonged cylindrical stigmas and the ovary.   C. The front view of an apical portion of the column showing the divergent thecae, the fleshy rostellum, the 2 stigmas and the orifice of the spur.   D. A pollinium on a caudicle attached to a viscidium.   E. The habit sketch of a fruiting branch showing the persistent bracts and fusiform capsules.

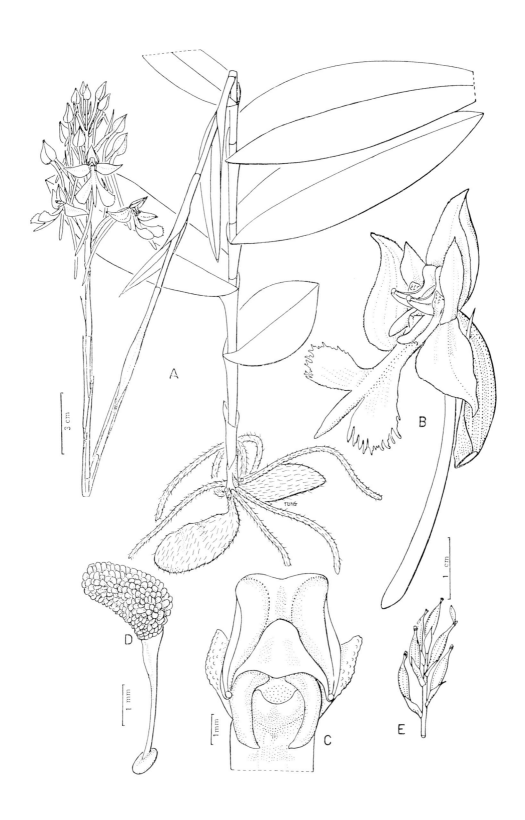

6. **Peristylus** Blume 潤蕊蘭屬 (FIGURE 10)

Flowers small, spirally arranged in a terminal spike; sepals subequal, free, erect, connivent; petals small, below the odd sepal and with it forming a hood; lip trilobed and calcarate, the spur shorter than the ovary, the disc fleshy, recurved, often with a mound-like callus, the lateral lobes entire, filiform or rarely ovate, the midlobe ovate, obtuse; column short, conic, auricles prominent, the apex entire, rotundate; rostellum yoke-shaped; anther 2-celled, the thecae parallel; pollinia 2, obovoid, sectile, the fragments relatively large, the caudicles short, equal or shorter than the pollinia; viscidia oblong or linear, relatively long; stigmas 2, convex, entirely adnate to the base of the lip and to the auricles of the column; ovary sessile. Capsules small, not much longer than the ovary, ellipsoid, with persistent perianth. Terrestrial orchids passing the dry cool winter in dormancy; with small oblong tubers on slender stalks. Leaves 2 or 3, close together, usually at the ground level, 2–12 cm. long. Five species native to Hong Kong, growing along paths in forested areas or on exposed slopes.

Key to the Species

A. Flowers white, crowded in fusiform spikes; lobes of lip equal, ovate-triangular; spur globose, rounded at the apex; leaves 3.5–5.5 cm. wide......*P. goodyeroides* (D. Don) Lindl.

AA. Flowers green or yellowish green, loosely spaced in slender spikes; lateral lobes of lip narrower than the middle lobe; leaves 1–3 cm. wide.

    B. Leaves ovate or ovate-elliptic, 2–4.5 cm. long; lateral lobes of lip ovate-falcate, 1 mm. long, 0.7 mm. wide; spur 4 mm. long, truncate at the apex..................................
    ....*P. densus* (Lindl.) Sant. & Kapadia

    BB. Leaves elliptic or oblong-elliptic, 3–10 cm. long; lateral lobes of the lip linear or filiform, narrower than the midlobe, 0.5 mm. or less wide; spur conic and pointed at the apex, or cylindric and bilobed or pointed at the apex.

        C. Spur conic, shorter than the lip, 2 mm. long, the apex pointed; leaves elliptic-oblong or elliptic-lanceolate.

            D. Leaves 2 or 3, elliptic-oblong, 2–7 cm. long, 1.5–2 cm. wide.....................
            ....(P. chloranthus Lindl.) *P. spiranthes* (Schauer) S. Y. Hu
            DD. Leaves 4 or 5, elliptic-lanceolate, 7–11 cm. long, 2–2.5 cm. wide...............
            ....*P. tentaculatus* (Lindl.) J. J. Sm.

        CC. Spur cylindric, equal or longer than the lip, the apex pointed or bilobed; leaves oblong-lanceolate, 5–10 cm. long, 1.5–2.5 cm. wide.................................
        ....*P. calcaratus* (Rolfe) S. Y. Hu

FIGURE 10. *Peristylus spiranthes*: A. The habit sketch of a flowering plant showing the ovate leaves at ground level, and the scape with many flowers. B_{1-2}. The lateral views of 2 flowers showing the bracts, the sepals, the petals, and the lip with oblong spur pointed at the apex, and short or filiform lateral lobes. C. The front view of a flower with portion of the ovary removed, showing the sepals, the petals, the lip with a mound-like callus at the center, the filiform lateral lobes turning inward, and a portion of the column with 2 thecae. D. The front view of a flower with all sepals and petals removed, showing the column with 2 parallel thecae, the folded rostellum, the lateral glands, and the 2 stigmas. E. The lateral view of the same showing the curved lip, the position of the lateral lobes, the oblong spur pointed at the apex, the rugous column with 2 lateral glands, and a stigma. F. A sectile pollinium with a short caudicle and a viscidium.

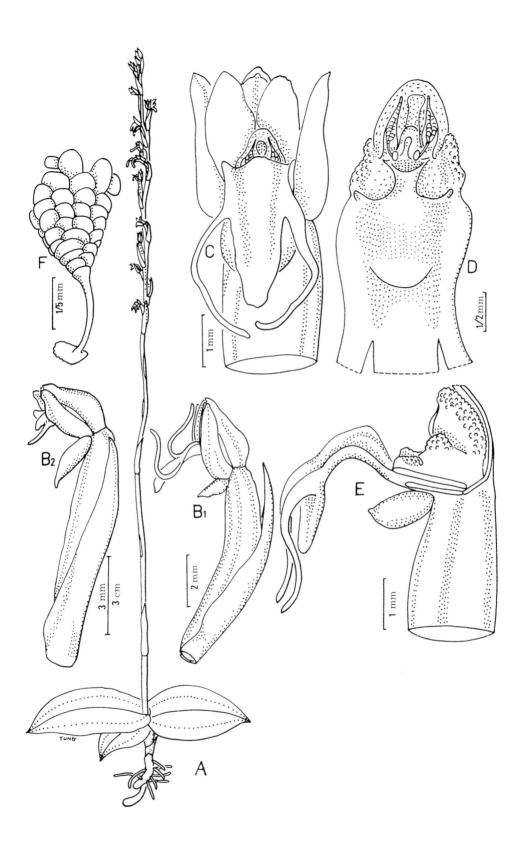

7. **Disperis** Swartz 雙袋蘭屬 (FIGURE 11)

Flowers small, 1 or 2, axillary to leafy bracts, terminal to a tiny stem; sepals spreading, the odd one erect, suborbicular, with the petals forming a shield behind the column, the lateral sepals connate at the base, separated at the middle, bent 90 degrees and calcarate; petals erect; column conic, obtuse; rostellum broad triangular, stipitate, with slender long and curved arms; anther 2-celled, the cells divergent; pollinia 2, small, sectile; caudicles elongated, the viscidia exposed; stigmas 2; ovary cylindric. Small succulent terrestrial orchids growing among grasses on arid peak of Lantau Island, overwinter in a dormant stage in form of small globose tubers. Stems slender, bearing 3 cataphylls and 2 tiny ovate cauline leaves. The species of the genus are predominantly African, with one species known from each of the following areas: Sri Lanka, Thailand, Java, The Philippines, Taiwan, Caroline Islands, New Guinea, and Queensland in Australia. One species in Hong Kong, *D. lantauensis* S. Y. Hu.

8. **Nervilia** Comm. 芋蘭屬 (FIGURE 12)

Flowers appearing before the leaves; sepals and petals similar, subequal, free, parallel to the column and slightly divergent at the apex; lip undivided, ecalcarate, the basal portion embracing the column, the apical portion subrotundate, slightly cleft, the disc inverted-carinate, and the base gibbous; column slender and long, clavate, thin and curved above the base; rostellum fleshy and truncate; anther terminal, subglobose; pollinia 4 in 2 pairs, finely granular, without caudicle or viscidium; stigma suborbicular; ovary short, campanulate, deeply sulcate, bent at the juncture with the pedicel. Terrestrial orchids requiring dormancy annually; rhizome slender, the end enlarged into a subglobose corm; erect stems short, subterranean, branched, each branch bearing a solitary leaf, cordate, lying flat over the ground. Scape delicate, aphyllous, bearing few scales and 2 flowers. One species in Hong Kong, *N. fordii* (Hance) Schlechter, very rare.

9. **Cryptostylis** R. Br. 隱柱蘭屬 (FIGURE 13)

Flowers non-resupinate; sepals subequal, spreading, involute and acuminate; petals shorter than the sepals; lip erect, undivided, ovate, acute, strongly concave at the base and surrounding the column, velvety-hairy, spotted; column very short, glandularly auriculate; anther erect, acute; pollinia 4, powdery; viscidium large, in direct contact with the stigma, without evident rostellum; stigma discoid, prominent, fleshy, occupying the entire front of the column; ovary cylindric. Capsules oblong, erect, erostrate. Terrestrial herb, glabrous. Roots thick, spreading, hairy. Rootstocks very short, erect. Leaves 2, petiolate, the petioles not sheathing; lamina elliptic, chartaceous. Scape simple, with imbricate cataphylls at the base. One species in Hong Kong, *C. arachnites* (Bl.) Hassk.; rare, growing on forest floor in ravines.

FIGURE 11. *Disperis lantauensis*: A. The habit sketch of a flowering plant showing a subglobose corm, and underground portion of the stem with cataphylls and a short aerial stem with one leaf and a single flower subtended by a leaf-like bract. B. The apical portion of a plant showing an abortive flower and a fully open flower with two sepals connate at base, saccate at middle and bending at an angle of 90°, the erect petals forming a shield-like structure with the dorsal sepal, the stipitate lip, folded and with horn-like lateral processes and a pendulous front lobe gradually attenuate and suddenly enlarged at the apex, the short column with 2 thecae, two pollinia, and a stipitate rostellum. C. The lip (FRONT) and the rostellum (BACK) with two lateral arms each bearing a viscidium.

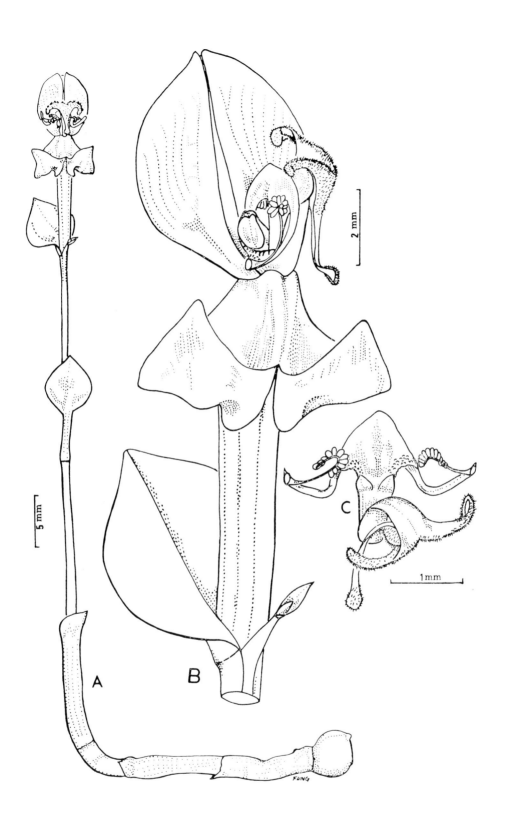

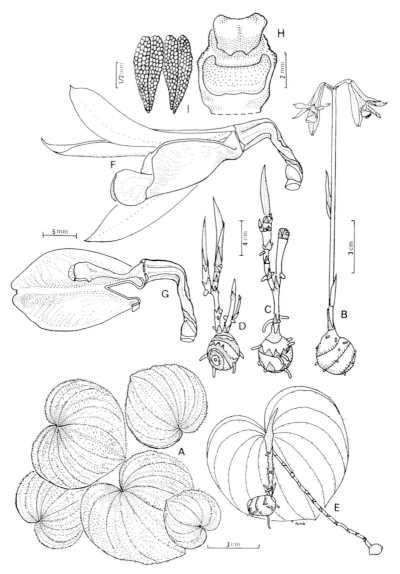

FIGURE 12. *Nervilia fordii*: A. The habit sketch of a colony of plants after anthesis, showing individual cordate leaves lying flat on the ground. B. The habit sketch of a flowering plant showing the globose corm, and the scape with cataphylls. C. The habit sketch of the subterranean portion of a plant just before the emergence of a leaf, showing a corm and the manner of branching after anthesis, indicating the first shoot giving rise to a scape bearing 2 flowers, and a second shoot bearing a leaf. D. The habit sketch of a corm showing the manner of development of leafy shoots, indicating the possibility of having 4 leaves. E. The habit sketch of a shoot emerged from a corm and bearing a leaf, and a long slender rhizome developed at its apical end and terminated by a cormlet, showing the vegetative means of multiplication of the species. F. The lateral view of a flower with a sepal and a petal removed, showing the lip attached to the base of the column, slightly gibbous at the base and folded at the apex. G. A flower with sepals and petals removed and the lip partially opened at base, showing the clavate column very thin near the base, the anther, and the stigma. H. The front view of the apex of the column showing the anther, the truncate rostellum and the stigma. I. Four slender pollinia in two pairs, without caudicle and viscidium.

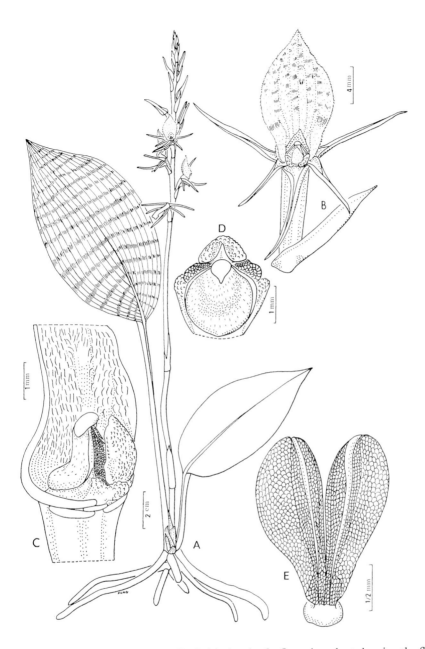

FIGURE 13. *Cryptostylis arachnites*: A. The habit sketch of a flowering plant showing the fleshy roots, the basal leaves, and the flowering scape, all emerged from and crowded at an erect short rootstock, and the non-resupinate flowers. B. The front view of a flower showing a bract, the ovary, and the relative positions of the spreading sepals, petals, and the uppermost ovate undivided lip. C. The lateral view of a flower with all sepals, and petals, and portion of the lip removed, showing the very stout column with lateral glands, the anther cap, the viscidium, one side and the front of a very large stigma, and the papillose inside of the basal portion of the lip. D. A front view of the column showing the anther cap, the viscidium, the very broad discoid stigma covering almost the entire front of the column, and the lateral glands. E. Four pollinia in two pairs.

10. **Spiranthes** L. C. Rich. 綬草屬 (Figure 14)

Flowers small, hardly open, in a terminal spiral spike; sepals subequal, free, delicate white; petals concealed, snow white or tinged pink; lip undivided, concave, the apex crisped, the base fleshy, locked to the base of the lateral sepals by a globose callus on each side, all the three attached to the apex of the ovary; column arcuate and slightly clavate; rostellum deltoid and acuminate, bifid, or merely a collar-like hyaline membrane attached to the middle of the pollinia, without an acuminate apex; anther shortly stipitate, acuminate, persistent; pollinia 4 in 2 pairs, powdery, obovoid and acute, with or without a viscidium; stigma orbicular or shield-like and tricuspidate along the upper margin; ovary oblong. Capsules oblong-ellipsoid, erect. Terrestrial perennial orchids. Roots fleshy, cylindric or fusiform. Rootstocks very short, erect, covered by the bases of the petioles and bearing small lateral buds. Two species in Hong Kong; both common.

Key to the Species

A. Rachis, bracts, sepals and ovary glabrous; stigma suborbicular; rostellum deltoid-acuminate, the apex bifid; viscidium elliptic............................*S. sinensis* (Pers.) Ames
AA. Rachis, bracts, sepals and ovary glandularly hairy; stigma shield-like, tricuspidate along the upper margin; rostellum narrow, collar-like, attached to the middle of the pollinia; without obvious viscidium...*S. hongkongensis* Hu & Barr.

11. **Manniella** Reichb. f. 滿氏蘭屬 (Figure 15)

Flowers small, inconspicuous, in a terminal hairy raceme; sepals subequal, the basal two-fifths connate and adnate to the upper portion of the ovary forming a pouch; petals slightly narrower, white; lip fleshy, undivided, concave, papillose and reflexed at the apex, the base unguiculate, with a hooked appendage on each side of the claw; column long, adnate to the perianth tube, the free portion terete, clinandrium with 2 teeth, one on each side of the anther; rostellum ovate, obtuse, unequally bilobed; anther oblong-ovoid, concave; pollinia 2, granular, folded on one side; viscidium terminal, separated; stigma 2, on the plane surface below the rostellum; ovary subcylindric, hairy. Perennial terrestrial herb, dormant in winter. Roots

---

Figure 14. *Spiranthes*: A. The habit sketch of a flowering plant of *S. sinensis*, showing a fascicle of fleshy roots, linear basal leaves, and a flowering scape with spirally arranged flowers in a spike. B–F. *Spiranthes hongkongensis*: B. The lateral view of a flower with glandular hairs on the ovary and sepals. C. The lateral view of a flower with dorsal sepal, lateral sepals, 1 petal, and one-half of the lip removed, showing a spherical callus on one side at the base and the hairs on the disc of the lip, the column with an anther above and the stigma beneath. D. The front view of an apical portion of the column showing the middle of two pollinia, a very narrow hyaline rostellum attached to the middle of the pollinia, and the shield-like stigma trilobed along the upper margin. E. Four pollinia in 2 pairs, without evident viscidium. F. The lateral view of the apical portion of the column showing the pollinia, the stigma, and the rostellum attached to the middle of the 2 larger pollinia. G–K. *Spiranthes sinensis*: G. The lateral view of a flower showing the glabrous ovary and sepals. H. The lateral view of a flower with the dorsal sepal, lateral sepals, 1 petal, and one-half of the lip removed, showing a spherical callus at the base and no hair on the disc of the lip, and the column with anther cap, pollinia, and stigma. I. The front view of the apical portion of the column showing the forked rostellum, the elliptic viscidium, and the oblong stigma. J. Four pollinia in 2 pairs attached to distinct elliptic viscidium. K. The lateral view of the apical portion of the column, showing the anther deep into the clinandrium, the rostellum bifurcate at apex, and the stigma.

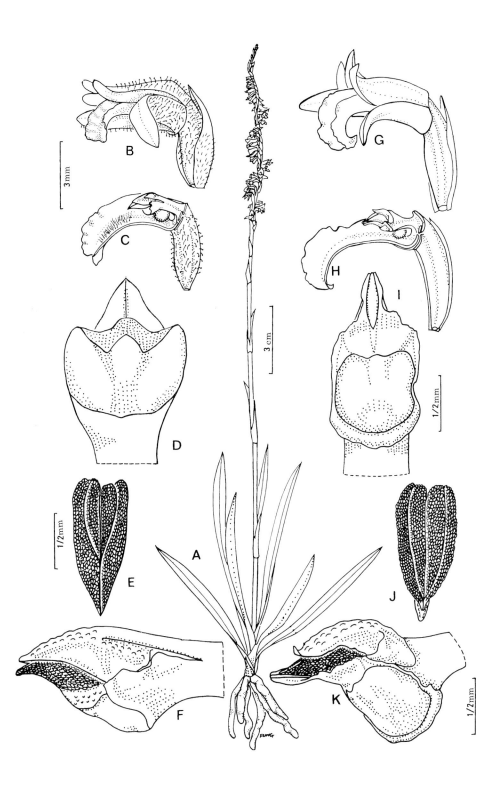

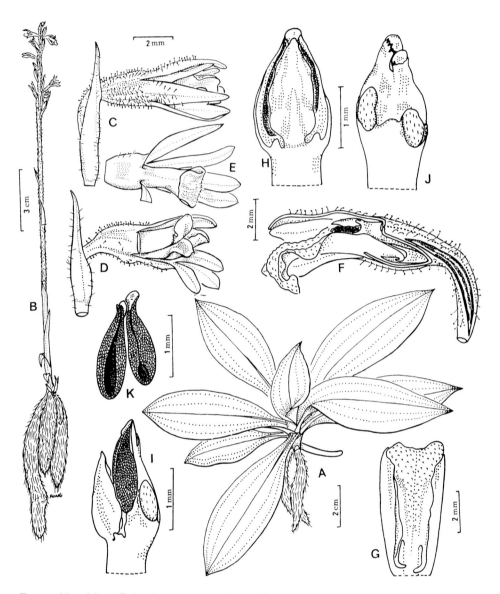

FIGURE 15. *Manniella hongkongensis*: A. The habit sketch of a plant with a rosette of leaves, and fusiform fleshy roots. B. The habit sketch of a flowering plant showing a scape with cataphylls, flesh fusiform roots, and a small vegetative bud at the side of the base of the scape. C. The lateral view of a flower with a bract showing the union of the basal 2/5 of the sepals to form a swollen tube. D. The lateral view of an abnormal flower with 1 sepal removed, showing the doubled condition of the lip. E. The ventral view of a portion of a flower showing the recurved tip of the lip. F. The lateral view of a flower with a petal, a lateral sepal and portion of the odd sepal removed, showing the fleshy lip with 2 hooked appendages in the pouch formed by the union of the basal portion of the lip and the 2 lateral sepals, the column with the anther slightly tipped up showing the 2 separate pollinia, and a portion of the ovary showing the parietal placentation. G. The adaxial view of the lip showing its union with the 2 lateral sepals, the 2 basal hooked appendages and the thick papillose disc. H. The dorsal view of the column showing the 2 apical lobes (representing staminodes), the clinandrium, the anther, the pollinia, the rostellum, and a large viscidium attached to the pollinium

42

fleshy. Rootstocks short, erect, bearing 2 buds, the flower bud unfolding before the foliar bud. Leaves all basal, oblong-elliptic, shortly petiolate. Scapes solitary, with sheathing basal scales. One species know from one locality in Hong Kong, *M. hongkongensis* Hu & Barr.

12. **Cheirostylis** Blume 叉柱蘭屬 (FIGURE 16)

Flowers small, inconspicuous, in a terminal small raceme; sepals connate, forming a tube above the middle; petals white, very small, mostly concealed; lip adnate to the sides or rarely to the base of the column, unguiculate, the claw with inflexed margin or simple and without folding margin, the apex dilated, often bilobed and fringed or dentate, rarely undivided and entire, the base furnished with 2 calli, rarely without callus; column short, the apex furnished with 2 glossy glands. and produced into an elongate appendage on each side; rostellum long and deeply cleft; anther ovate-acuminate, deeply sunk in the clinandrium; pollinia 4 in 2 pairs, sectile, with or without slender caudicle; viscidium oblong or obovate, cuneate; stigmas 2, connivent to the lateral glands; ovary ovoid or oblanceolate. Capsules obovoid-oblong, erostrate. Epiphytic delicate perennial orchids with worm-like succulent short rhizomes, hairy on the contact surface, clinging to wet boulders below trees along streams. Leaves basal, petiolate, ovate or elliptic. Four species in Hong Kong.

Key to the Species

A. Ovary and calyx tube hairy.
    B. Lip fringed; hairs on ovary and calyx glandular; calli consisting of 2 rows of teeth...
    ....*C. chinensis* Rolfe
    BB. Lip dentate; hairs on ovary and calyx not glandular; calli tricuspidate.................
    ....*C. jamesleungiana* Hu & Barr.
AA. Ovary and calyx tube glabrous.
    B. Lip bilobed and fringed at the apex; base of the lip subsaccate, with two bifid and
    horn-like calli.... ....*C. monteiroi* Hu & Barr.
    BB. Lip simple, entire; base of lip unguiculate, without any calli............................
    ....*C. clibborn-dyeri* Hu & Barr.

13. **Anoectochilus** Blume 開唇蘭屬 (FIGURE 17)

Flowers medium-sized, in a loose terminal raceme; sepals free, at a right angle with the ovary at anthesis, the odd one concave and strongly recurved at the apex, the lateral ones obliquely auriculate at the base; petals connivent with the odd sepal and with it forming a recurved galea concealing the column; lip fleshy, adnate to the base of the column, calcarate, bending with a sharp angle, prolonged into a narrow channelled and fringed claw with infolding and meeting margins, widened transversely and 2-lobed at the apex, the spur conic, bilobed at the tip, with 2 large cristate calli; column short, with 2 lateral glands, the apex winged and auriculate, clinanthium cyathiform; rostellum elongated, deeply bifid; anther suberect, ovate,

---

of the right side. I. The lateral view of the column showing the anther, a staminode, 2 pollinia, a beak-like rostellum forked at the apex, and a stigma. J. The ventral view of the column showing the rostellum unequally lobed, each lobe with a viscidium, and the 2 lateral stigma at uneven levels. K. Dorsal view of the 2 pollinia separated, folded, each with a viscidium.

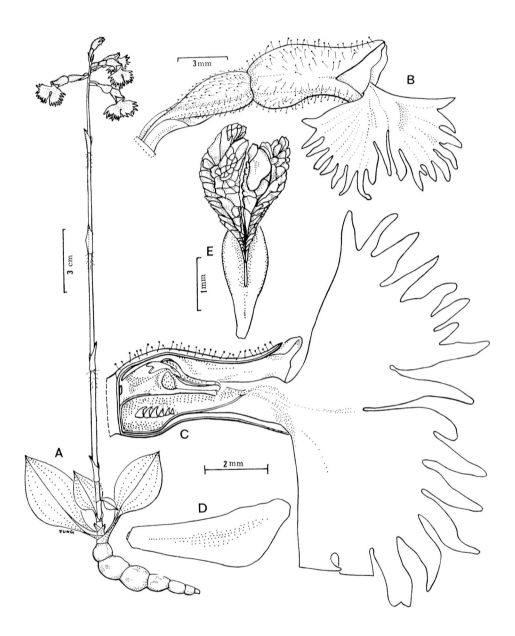

FIGURE 16. *Cheirostylis chinensis*: A. The habit sketch of a flowering plant showing the worm-like succulent rhizome, the short leafy shoot terminated by a scape bearing 5 flowers. B. The lateral view of a flower showing the bract, the ovary and the urceolate calyx, both glandularly hairy, the 2 petals, and the extraordinarily dilated bilobed lip fringed along the margin. C. The lateral view of a flower with half of the calyx, half of the basal portion of the lip, and 1 petal removed, showing the slightly saccate base of the lip attached to the side of the column and with a row of 5–6 tubercles on one side, the folded margin of the side, and the suddenly expanded apical lobe, the short fleshy column with a lateral gland associated with a stigma, the anther and one side of a long arm of the rostellum. D. An oblanceolate petal obliquely truncate at the apex. E. Four pollinia in 2 pairs attached to an elliptic viscidium.

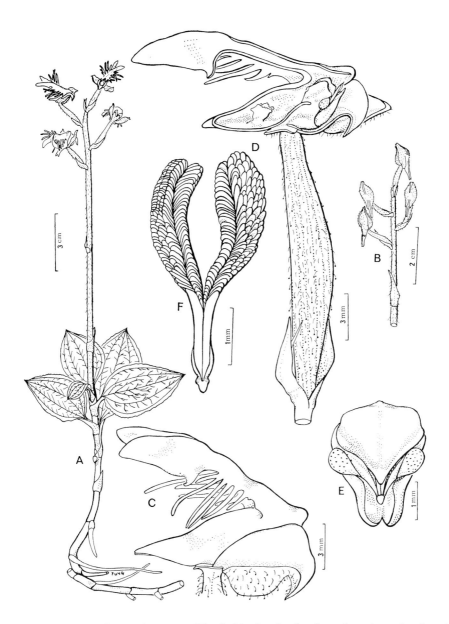

Figure 17. *Anoectochilus yungianus*: A. The habit sketch of a flowering plant showing the slender rhizome, the ascending stem with 5 leaves, and a scape terminated by a 4-flowered raceme. B. The apical portion of a scape with mature flower buds. C. The lateral view of the upper portion of a flower with a lateral sepal removed, showing the dorsal sepal and the petal strongly reflexed at the apex, and the lip with bilobed conic spur, and a 90° bending, with the channeled claw infolding and fringed, and the 2 enlarged apical lobes. D. The lateral view of a flower with bract and ovary, with half of the dorsal sepal, 1 lateral sepal, 1 petal, and half of the lip removed, showing the column with 2 fleshy auricles in front, the 2 crested glands inside the spur, the infolding margin of the channeled claw, the anther, the pointed rostellum and a lateral stigma. E. The apical portion of the column showing the anther, the stigmas, and the rostellum above the auricles. F. Two pollinia with long caudicles and viscidium.

acuminate; pollinia 2, sectile; caudicles slender; viscidium oblong, partially exposed; stigmas 2, below the rostellum; ovary cylindric. Capsules oblong, erect, erostate, with acute ridges. Terrestrial perennial orchids with creeping rhizomes and fibrous roots. Leaves ovate, petiolate, crowded at the ground level, variegated along the veins. One species in Hong Kong, *A. yungianus* S. Y. Hu, growing on moist forest floor.

14. **Ludisia** A. Rich. 血蝶蘭屬 (FIGURE 18)

Flowers medium-sized, in a terminal raceme; sepals subequal, free, the odd one erect, ovate, concave, the lateral ones spreading backward; petals connivent with the odd sepal and with it forming a galea; lip adnate to the base of the column, saccate, twisted, grooved, widened and divergently bilobed at the apex, the sac containing 2 stalked calli; column subterete, twisted toward the opposite direction from the lip, enlarged at the apical end; clinandrium concave; rostellum bifid, the lobes unequal, twisted; anther semi-immersed, acuminate; pollinia 2, sectile, caudicles slender; viscidium oblong, partially exposed, stigma 1, on the plane surface in front of the column; ovary cylindric, hairy. Capsules oblong, erect. Terrestrial perennial orchids with creeping fleshy rhizomes and fibrous roots. Leaves cauline, petiolate, ovate, elliptic, reddish brown with white major veins. One species in Hong Kong, *L. discolor* (Ker-Gawl.) A. Rich. (*Haemaria discolor* [Ker-Gawl.] Lindl.).

15. **Goodyera** R. Br. 斑葉蘭屬 (FIGURE 19)

Flowers small or medium-sized, in a terminal spike; sepals subequal, the odd one ovate, concave, the lateral ones spreading sidewise or backward; petals connivent with the odd sepal and with it forming a hood; lip adnate to the base or basal sides of the column, undivided, concave, ventricose, or subsaccate, fimbriate or villose on the inside, recurved at the apex; column subterete, short; clinandrium dilatate, the margin continued with the rostellum forming a cup; rostellum long, divided into 2 arms; anther ovate, acuminate; pollinia 4 in 2 pairs, sectile, oblong-acuminate; caudicles slender; viscidium oblong or linear; stigma concave, large; ovary subterete, striate-sulcate. Capsules erect, obovoid or oblong, erostrate, with persistent perianth at the apex. Terrestrial orchids with creeping or suberect rhizomes, and fibrous roots. Leaves cauline, petiolate, ovate or ovate-elliptic. Four species in Hong Kong.

Key to the Species

A. Rhizomes suberect; leaves elliptic, 7–12 cm. long, 2–5 cm. wide; scapes 12–25 cm. long; flowers 30 or more, 2–3 mm. across, inconspicuous, white.....*G. procera* (Ker-Gawl.)Hook
AA. Rhizomes creeping; leaves ovate-elliptic, 3–6 cm. long, 1.5–3.5 cm. wide; scapes 2–10 cm. long; flowers 1–15, 8–25 mm. across.
    B. Lateral sepals spreading sidewise; ovary and sepals glandularly hairy; lip adnate to the basal sides of the column..............................*G. foliosa* (Lindl.) Benth. ex HK. f.
    BB. Lateral sepals spreading backward; ovary and sepals without glandular hairs; lip attached to the base of the column.
        C. Flowers 4 or 5 to a scape; lateral sepals 4 mm. wide; stigma concealed by high margin of a pouch; leaves grayish brown.....*G. cordata* (Lindl.) Benth. ex HK. f.
        CC. Flowers 1 or 2 to a scape; lateral sepals 7 mm. wide; stigma subtruncate, not concealed by high margin; leaves shiny green..............*G. youngsayei* Hu & Barr.

16. **Zeuxine** Lindl. 線柱蘭屬 (FIGURE 20)

Flowers small, inconspicuous, hardly opening, in a terminal spike; sepals subequal, the odd one concave and gibbous at the base; the lateral sepals oblique

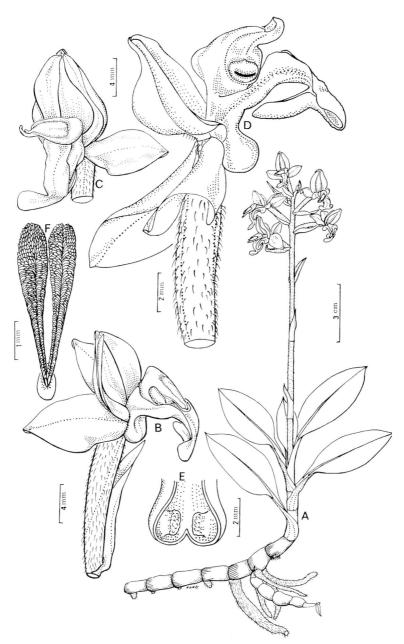

FIGURE 18. *Ludisia discolor*: A. The habit sketch of a flowering plant showing the fleshy root, the creeping succulent rhizome with an off-shoot, and the ascending leafy stem terminated by a scape. B. The lateral view of a flower showing the bract, the relative position of sepals and petals, the saccate lip partially covering the column at the base, twisted and enlarged into 2 lobes at the apex, and the twisted column with the stigma on the left side and the anther partially visible on the right side. C. A flower as seen from above, showing the twisted lip, the column, the anther, and the unequal and slightly twisted rostellum-lobes. D. A flower as seen from beneath, showing the saccate lip, the unequally lobed rostellum with the viscidium in the curve of one lobe, and the stigma. E. The spur of the lip split open showing the 2 glands. F. Two pollinia attached to the viscidium.

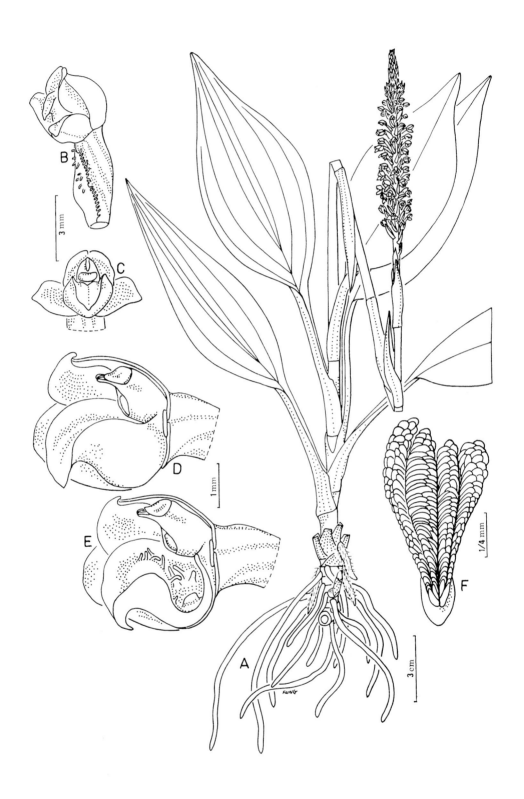

ovate; petals small, forming a hood with the odd sepal; lip fleshy, adnate to the sides of the column, parallel to the sepals and petals, fleshy, saccate at the base, bilobed or entire at the apex, the sac containing 2 calli, the apex transversely widened, often bilobed, connected to the saccate base by an elongated fleshy claw with inflexed margin; column short, stout, with 2 apical glands; rostellum large, deeply divided; clinandrium membranaceous, dilated, cupular; anther ovate-oblong, acute; pollinia 2, sectile, attached to a spathulate or linear-oblanceolate stipe; viscidium subdiscoid; stigmas 2, below the rostellum; ovary cylindric or ellipsoid, striate-sulcate. Capsules ellipsoid, striate-sulcate. Terrestrial orchids. Leaves cauline. Four species in Hong Kong.

Key to the Species

A. Leaves ovate, crowded together; apical portion of the lip with divergent lobes; calli bifid at the apex, or lunato-ligulate.
    B. Lip white, yellow at the throat; calli bifid at the apex........*Z. gracilis* (Breda) Blume
    BB. Lip bright yellow; calli lunato-ligulate...............................*Z. leucochila* Schlechter
AA. Leaves linear-oblong or lanceolate, evenly disposed on the stem; apical portion of lip entire or with 2 parallel lobes; calli cushion-like.
    B. Apical portion of lip bilobed; scapes, bracts, sepals and ovary glabrous; leaves linear-oblong, obtuse at the apex...*Z. strateumatica* (L.) Schltr.
    BB. Apical portion of lip obovate; scapes, bracts, sepals and ovary villose; leaves lanceolate, acute at the apex...*Z. membranacea* Lindl.

17. **Hetaeria** Blume 伴蘭屬 (FIGURE 21)

Flowers non-resupinate, small, hardly opening, in a terminal raceme; sepals free, subequal, ovate; petals largely concealed; lip uppermost, adnate to the sides of the column, fleshy, undivided, strongly concave, the base furnished with papillose calli, the apex short and inflexed; column short, with 2 lateral glands, the apex with fleshy appendages and adnate to the lip; rostellum erect, elongated and bilobed; anther erect, ovate, acute; pollinia 2 or 4 in 2 pairs, sectile, obovoid and acuminate; caudicles slender; viscidium oblong, affixed between the lobes of the rostellum; stigma transversely 2 lobed, below the erect rostellum; ovary oblong-cylindric, erect, striate-sulcate. Capsules oblong-ovoid, erect. Terrestrial orchids with creeping rhizomes. Leaves cauline, ovate, petiolate. Two species in Hong Kong.

Key to the Species

A. Plants 40 cm. or more high; sepals one-half as long as the ovary, glandularly hairy; lip ventricose, the calli papillose; appendage of the column dilated and reflexed at the apex; lip acute at the apex...*H. nitida* Ridley
AA. Plants 20 cm. or less high; sepals one-third as long as the ovary, pilose but not glandular; lip saccate, the calli subglobose and stipitate; appendage of the column dilated at the base; lip truncate at the apex and folded...*H. cristata* Blume

---

FIGURE 19. *Goodyera procera*: A. The habit sketch of a flowering plant showing the roots, the rhizome, and the stem terminated by a scape. B. The lateral view of a flower showing the bract, the ovary, and the perianth. C. The front view of a flower showing the lateral sepals, 2 petals, the lip, the forked rostellum, the viscidium, and the stigma. D. The lateral view of a flower with half of the dorsal sepal, 1 lateral sepal, and 1 petal removed, showing the ventricose lip, the obovoid column, the anther, the rostellum, and the stigma. E. The same with half of the lip removed, showing the column, the fleshy thick disc of the lip, a callus, and the hairs. F. Four pollinia in 2 pairs, attached to a viscidium.

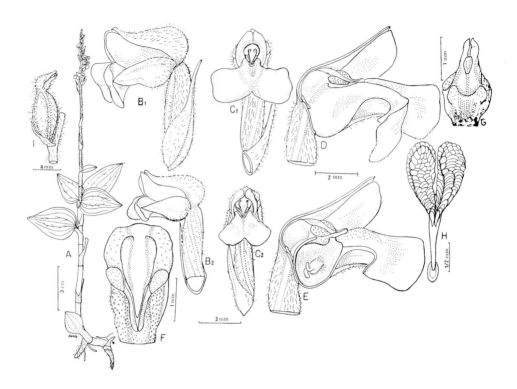

FIGURE 20. *Zeuxine gracilis*: A. The habit sketch of a flowering plant. B_1–C_1. The lateral and front views of a flower of a plant from the New Territories. B_2–C_2. The lateral and front views of a flower of a plant from Hong Kong Island. D. The lateral view of a flower with the lateral sepal, 1 petal and one-half of the dorsal sepal removed, showing the saccate lip adnate to the short column and the lateral glands associated with the stigma. E. The same with one-half of the lip removed, showing the glands at the base of the spur. F. The top front view of the column showing the anther, the lateral glands and the rostellum. G. The front view of apical portion of column with the rostellum pushed back, showing lateral glands, the viscidium, and the stigmas below the rostellum. H. The pollinia on a linear lanceolate stipe and the subdiscoid viscidium. I. An ellipsoid fruit with the persistent sepals.

FIGURE 21. *Hetaeria nitida*: A. The habit sketch of a flowering plant showing 5 years' growth, with the older stems lying over the soil as rhizomes. B. The lateral view of a flower showing the non-resupinate position, with the odd sepal lowermost. C. The front view of a flower showing the relative positions of the 3 sepals, the 2 large ovate petals, the fleshy papillose lip acute at the apex and folded along the sides, the column with 2 conspicuous glands, one on each side, the 2 broad arms of the rostellum and the anther. D. The lateral view of a flower with 1 lateral sepal, half of the odd sepal and 1 petal removed, showing the fleshy papillose lip, the very short column, a large gland, an arm of the rostellum, and the anther. E. The same with half of the lip removed, showing a group of 6 papilli on one side of the base of the thick fleshy lip, the very short column extending forward into 2 broad appendages papillose on the inside and recurved along the margin, attached to the lip at the lower half, 1 large conspicuous gland, 2 arms of the rostellum, and the anther. F. The front view of the column showing, from front to back, the papillose appendages curved outward, the stigmas, the rostellum with 2 broad arms, and the oblong viscidium. G. Four pollinia in 2 pairs, on a slender stipe attached to an oblong viscidium.

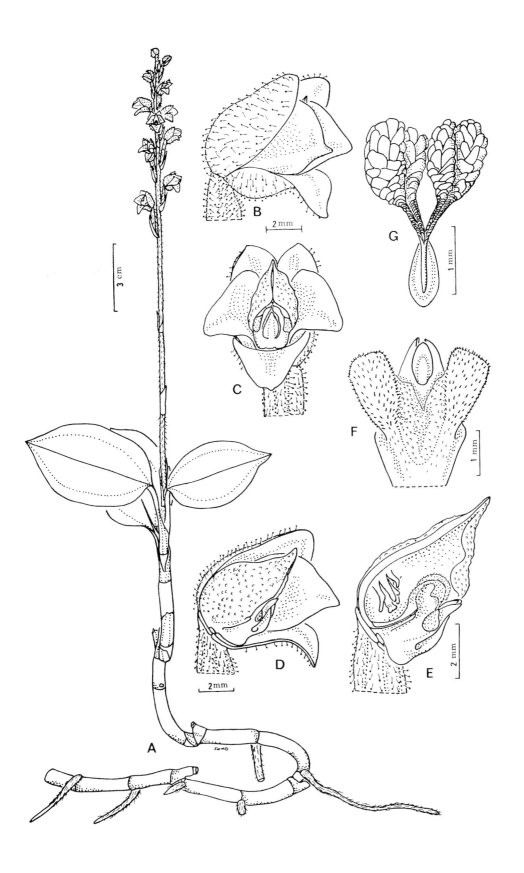

51

18. **Vrydagzynea** Blume 二尾蘭屬 (FIGURE 22)

Flowers small, non-resupinate, hardly open, in a terminal spike; sepals subequal, free, the odd one ovate, concave, the lateral ones obliquely auriculate at the base; petals smaller than the sepals and concealed by them; lip erect, undivided, spurred, fleshy, concave, the spur oblong-conic, with two hanging stipitate calli; column very short, stout; rostellum long, bifurcate; anther erect, ovate; pollinia 2, sectile, obovoid and acuminate; caudicles short; viscidium obovoid, between the arms of the rostellum; stigmas 2, below the rostellum; ovary oblong, striate-sulcate. Capsules oblong, erect, erostrate. Terrestrial orchids with creeping rhizomes. Erect stems leafy. Leaves ovate-elliptic, at the ground level, forming a rosette. Scapes with foliaceous bracts. One species in Hong Kong, *V. nuda* Blume, growing on damp forest floor, rare.

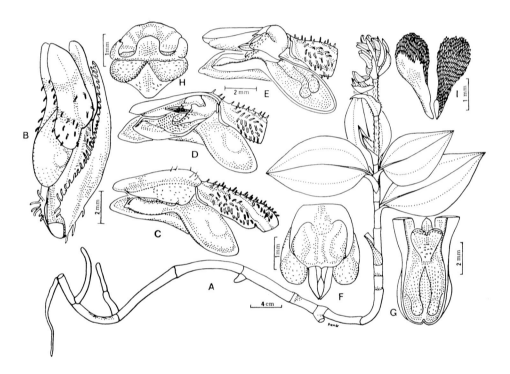

FIGURE 22. *Vrydagzynea nuda*: A. The habit sketch of a flowering plant showing the long, creeping rhizome, and the ascending leafy stem, terminated with a short raceme. B. The lateral view of a flower subtended by a bract, showing the sepals very close together at anthesis, and a large saccate spur. C. A flower with a sepal removed, showing an oblique petal and the lip. D. The same with portion of the odd sepal, 1 lateral sepal, and 1 petal removed, showing the short column, the anther, the caudicle of the pollinia, a viscidium, the edge of the rostellum, 1 stigma with glandular side, and the midlobe of the lip projecting forward at the center and rolled in along the sides. E. The same with half of the lip removed, showing the thick front disc of the lip, 2 stigmas, the rostellum, and 2 stalked glands in the spur. F. The top view of the column showing the anther cap, 2 discoid stigmas, the caudicles of the pollinia, the viscidium, and a narrow portion of the rostellum. G. The ventral view of the column and portion of the spur showing the tip of the rostellum, the glandular nature of the tissue below the stigmas, and the 2 stalked glands in the spur. H. Front view of the column with the rostellum pushed back a little, showing the shape of the stigmas. I. Dorsal view of a pollinium (RIGHT), and the ventral view of the same (LEFT).

19. **Tropidia** Lindl. 竹莖蘭屬 (FIGURE 23)

Flowers in a short axillary spike; sepals narrow, subequal, the lateral ones broader than the odd one, connate at the base and enclosing the lip; petals subsimilar to the odd sepal; lip undivided, equal to the sepals in length, caniculate, embracing the column, shortly calcarate, acuminate at the apex; column short, bearing the anther on the back; anther erect, ovate-acuminate; pollinia 2, caudicles subulate, viscidium peltate; rostellum elongate, the sides dilated and membranous partially covering the anther, the apex subulate, bifid; ovary oblong. Capsules ovoid-oblong. Terrestrial orchids with bamboo-like habit, the stems foliaceous. One species in Hong Kong, *T. hongkongensis* Rolfe, vary rare.

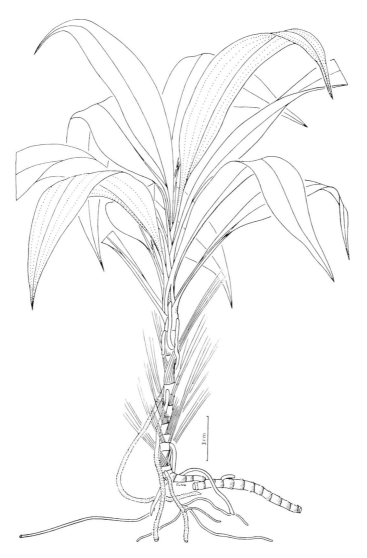

FIGURE 23. *Tropidia hongkongensis*: The habit sketch of a plant showing the roots, the cane-like rhizome, the stiff stem with the fiber of disintegrated leaves, the aerial roots, and the 6–12 cauline leaves with long petioles, plicate-venose, ovate-oblong to lanceolate blades.

20. **Aphyllorchis** Blume 無葉蘭屬 (Figure 24)

Flowers small, in a terminal raceme; sepals subequal, similar, slightly spreading, rather fleshy, oblong, subcymbiform, obtuse at the apex, with a median longitudinal ridge on the back; petals linear-oblong, obtuse, slightly turning outward at the apical end, auriculate at the base, with a median longitudinal ridge; lip 3-lobed, the lateral lobes basal, triangular, acute, erect; the midlobe fleshy, ovate, attached to the glandular and rugose disc elastically, relatively large, concave, the margin wavy and crenate, the apex obtuse; column slender, cylindric, curved, fleshy, smooth, without wing or any other decoration, the apex truncate; clinandrium truncate, cyathiform; anther fleshy, subglobose-ovate, concave, thecae lateral, parallel; pollinia 4 in 2 ellipsoid pairs, powdery, often affixed to the rostellum; rostellum narrow, fleshy, hardily distinguishable from the stigma; stigma large and wide, obovoate, the upper margin subtrilobed; ovary cylindric, striate, slightly rugose. Capsules subfusiform, pendulous, strongly rostrate. Saprophytic or parasitic, aphyllous. Roots fleshy, cord-like. Rhizomes short, horizontal, with membranous black scales, buds inconspicuous, subterminal, axillary to a scale. Cataphylls white, short, sheathing. Stem 30–50 cm. tall, terminated by a loose raceme; bracts membranous, reflexed, parallel to the rachis, purple. Flowers yellow; ovary violet-purple. One species, recently reported from the hills back of Shatin, *A. montana* Reichb. f.; appearing in flower in August–September, growing in shade of large trees.

21. **Cephalantheropsis** 嶺南黃蘭屬 (Figure 25)

Flowers medium-sized, showy, in lateral racemes; sepals and petals subsimilar, free, spreading; lip attached to the base of the column, 3-lobed, the lateral lobes erect and the apex recurved, midlobe transversely widened, 2-lobed, crisped; column suberect, short, stout, winged on the sides, slightly enlarged at the base, the apex truncate; clinandrium narrow; rostellum ovate, acute; anther ovoid, incumbent, hanging on the upper margin of the clinandrium; pollinia 8, obovoid-acuminate; viscidium small, discoid; stigma suborbicular; ovary cylindric, striate-sulcate. Capsule oblong, rostrate. Subepiphytic perennials growing in shade along rocky bed of streams, with the roots clinging to a wet rock. Stem bamboo-like, 2 or 3 together, leafy. Leaves all cauline, chartaceous, plicate-venose, disarticulate below the lamina. Scapes emerging from the axils of cataphylls or lower leaves, with basal imbricate scale. One species in Hong Kong, *Cephalantheropsis gracilis* (Lindl.) S. Y. Hu.

FIGURE 24. *Aphyllorchis montana*: A. The habit sketch of a flowering plant showing the roots on a short rhizome and the aphyllous aerial shoot with a terminal raceme. B. The rhizome and roots of the same, showing an axillary bud below the emergency of the aerial shoot. C. A flowering shoot showing two flowers and several buds all subtended by reflexed bracts. D. The front view of a flower showing 3 sepals, 2 petals, a trilobed lip with triangular lateral lobes and an ovate midlobe wavy along the margin, rugose at the disc, the column, anther with lateral thecae, the narrow rostellum and a large stigma. E. An abnormal flower with all three petals lip-like, each with 2 triangular lateral basal lobes and an ovate wavy midlobe, and glandular disc, the slender column with punctiform stigma and without anther. F. The lateral view of a flower with one-half of the odd sepal, the lateral sepals, and one-half of the lip removed, showing the petal auriculate at the base and ridged on the back, the lip with a triangular lateral lobe, an ovate midlobe, and a rugose disc, the cylindric column, the terminal anther with 2 lateral thecae, and a large stigma. G. The lateral view of the curved column, with a subglobose terminal anther, and a large stigma. H. Two ellipsoid pairs of pollinia without caudicle or stipe. I. A young fruit with persistent perianth and column.

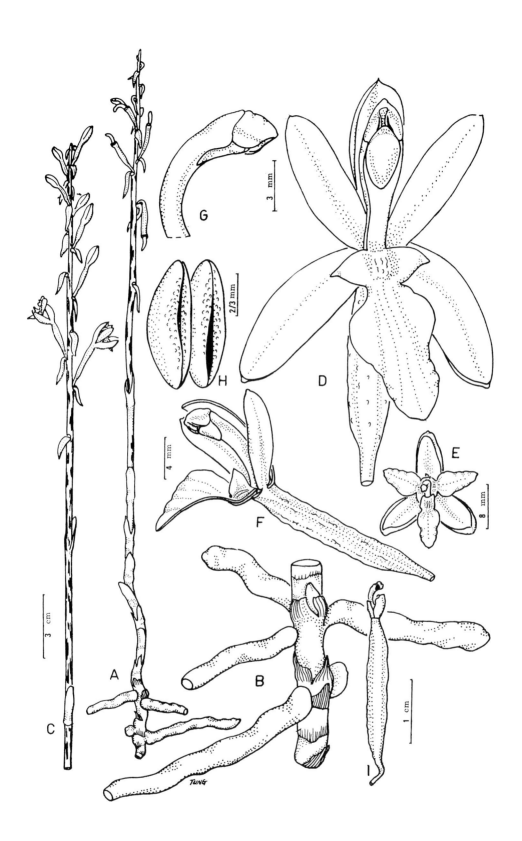

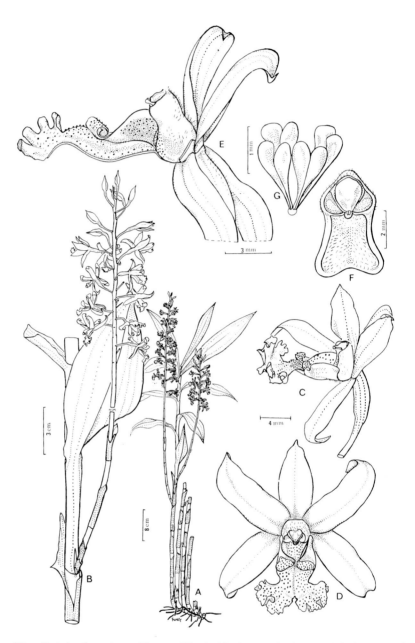

FIGURE 25. *Cephalantheropsis gracilis*: A. The habit sketch of a 7-year old plant showing the bamboo-like habit with persistent old stems, and the flowering shoot bearing 2 lateral racemes below the leafy portion. B. A portion of the same consisting of the upper inflorescence and 1 leaf, showing the flowers with the bracts dropped off. C. The lateral view of a flower showing the sepals, petals, column, and the lip with recurved lateral lobes, and midlobe with ruffled edge. D. The front view of a flower showing the symmetrical arrangement of the sepals and petals, the rolled lateral lobes of the lip, and the midlobe with a deep notch. E. The lateral view of a flower with a petal, the lower portion of 2 sepals and half of the lip removed, showing the attachment of the lip to the base of the column. F. The column showing the position of the anther and stigma. G. Eight obovoid pollinia on short caudicles and a viscidium.

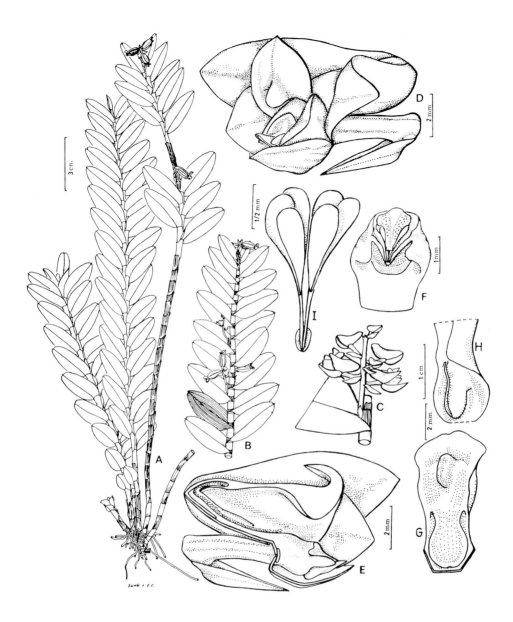

FIGURE 26. *Appendicula bifaria*:   A. The habit sketch of a fruiting plant showing five years' growth, with the fruit on the fourth year's growth. The fifth year's growth is leafless.   B. A fruiting branch showing the axillary position of the inflorescence.   C. An inflorescence with a basal scale, 2 non-resupinate flowers, and 4 buds.   D. A flower enlarged showing the bract, the odd sepal at an inferior positions, the lateral sepals, the lip at a superior position with a horizontal spur, and a unicorn-like callus at the center of the midlobe.   E. The lateral view of a flower with half of the odd sepal, 1 petal, 1 lateral sepal and half of the lip removed, showing the short column with 2 cuspidate arms, and the anther, and the basal appendage of the lip. F. The front view of a column showing the anther, pollinia, rostellum, and the stigma.   G. The top view of lip showing the unicorn-like callus at the bend, and the rounded concave appendage at the base.   H. The lateral view of basal portion of the lip showing the attachment of the rounded concave appendage.   I. Six pollinia in 2 groups on a forked stipe.

22. **Appendicula** Blume 牛齒蘭屬 (FIGURE 26)

Flowers very small, non-resupinate, in axillary spikes; sepals subequal, the odd one ovate, concave, the lateral ones adnate to the column-foot forming a mentum; petals ovate, obtuse; lip fleshy, undivided, attached to the column-foot and produced into a conic spur filled with a pendulous appendage, the apex reflexed, disc with an unicorn-like callus; column short, stout, 2-cuspidate at the apex; clinandrium small; rostellum triangular, long acuminate, 2-lobed; anther sessile, acuminate; pollinia 6, in two groups of 3, attached to a slender caudicle; viscidium discoid; stigma suborbicular; ovary cylindric, subsessile. Capsules oblong-ellipsoid, with persistent perianth at the apex. Epiphytic caespitose perennial orchids. Stems erect. Leaves numerous, distichous. Spikes several on the same shoot, axillary, with imbricate scales. One species in Hong Kong, *A. bifaria* Lindl. ex Benth., growing on wet rocks in the shade of trees along stream beds.

23. **Dendrobium** Swartz 石斛屬 (FIGURES 27–28)

Flowers showy, resupinate, medium-sized or rather small, in loose racemes, or occasionally solitary, lateral or slightly below the shoot apex, with imbricate basal scales; sepals rather fleshy, unequal, the odd one symmetrial, ovate, the lateral sepals oblique, adnate to the column-foot forming a mentum; petals thinner and often broader, rarely narrower than the sepals; lip undivided, attached to the apex of the column-foot, narrowed at the base, the sides rising up and usually embracing the column, the apical portion enlarged and colorful, disc lamellate; column short, stout, the base extended into an elongated foot, and the apex winged; clinandrium produced into a tooth on each side; rostellum broad, the upper margin recurved, obtuse, and the lower margin retused; anther terminal, incumbent, convex, often 2-lobed and papillose; pollinia 4 in 2 pairs, waxy, ovoid or oblong, slightly compressed laterally, without caudicle and viscidium; stigma suborbicular; ovary obovoid or subcylindric, the basal end curved downward. Capsules oblong or obovoid. Epiphytic perennial orchids. Roots numerous, elongated, clinging to the substratum. Stems leafy, erect or pendulous, emerging from the base of a shoot giving a caespitose appearance, or rarely from an upper node. Leaves oblong, elliptic or bilaterally compressed. Flowers appearing on leafy shoots or often after the leaves dropped off. Many species cultivated for medicinal uses or for ornamental purposes. The species included in the following key are all used in Chinese medicine.

Key to the Species

A. Leaves flat, with obvious upper and lower surfaces; flowers 2–6 cm. across; sides of the lip covering the column entirely or partially; mentum 2–3 times as long as the column.
    B. Apical margin of the lip fringed.
        C. Flowers purplish-pink; stems less than 20 cm. long, 5 mm. in diameter; leaves 2–3.5 cm. long...*D. loddigesii* Rolfe
        CC. Flowers yellow; stems 30–100 cm. long, 1 cm. or more in diameter; leaves 10 cm. or more long...*D. fimbriatum* Hooker
    BB. Apical margin of the lip not fringed.
        C. Petals and lip wavy and crisped, rounded at the apex; flowers 5 cm. across; leaves oblong-linear, obtuse at the apex................................*D. nobile* Lindl.
        CC. Petals and lip smooth or denticulate, acute at the apex; leaves elliptic, acute at the apex.

D. Flowers 5 cm. or more across; margin of the lip denticulate; mentum shorter
   than the column..*D. linawianum* Reichb. f.
DD. Flowers 4 cm. or less across; margin of the lip entire; mentum 2–3 times as
    long as the column.
     E. Lip ventricose; flowers purplish-pink, 1–3 in a cluster; leaves 8–10 cm.
        long...*D. hercoglossum* Reichb. f.
     EE. Lip oblanceolate; flowers 3–5 in a cluster, white; leaves 3–6 cm. long...
        ....*D. moniliforme* (L.) Sw.
AA. Leaves bilaterally compressed, without obvious upper and lower surfaces; flowers less
    than 1 cm. across; sides of the lip not covering the column; mentum 5 times longer than the
    column...*D. acinaciforme* Roxb.

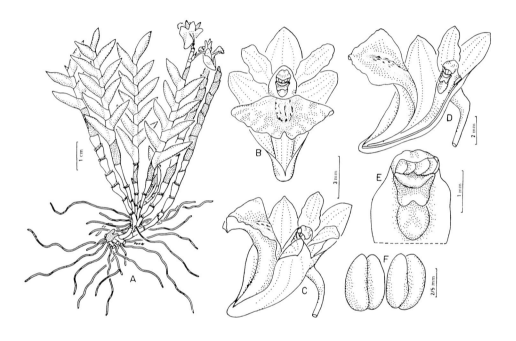

FIGURE 27. *Dendrobium acinaciforme*: A. The habit sketch of the plant showing the caespitose
habit, the numerous string-like roots, and the upright stems (4 leafy, 2 leafless, 1 flowering,
and a new shoot). B. The front view of a flower showing the wide sepals, the narrower petals,
the fan-like lip, and the column. C. The lateral view of a flower showing the lateral sepals
adnate to the column foot to form a mentum, the lip, the column, and the short ovary. D.
The same with 1 lateral sepal, and 1 petal removed, showing the column with prolonged foot,
the mentum, and the lip. E. The front view of the column showing the anther, the pollinia,
the broad rostellum recurved along the upper margin and notched at lower portion, and the
stigma. F. Four pollinia in 2 pairs without caudicle and viscidium.

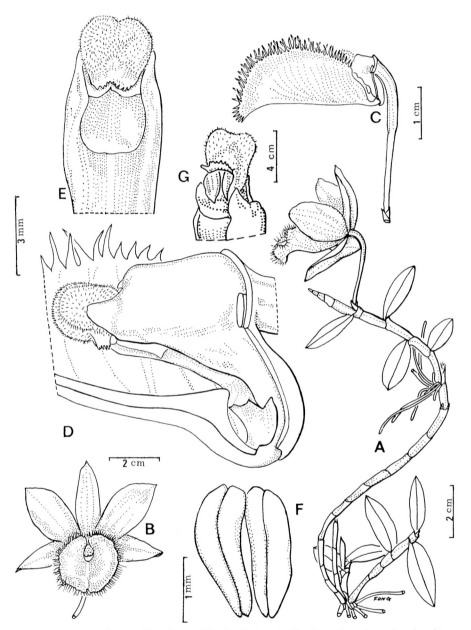

Figure 28. *Dendrobium loddigesii*: A. The habit sketch of a flowering plant showing 2 young shoots and a flowering shoot with aerial roots emerging from a leafless stem. B. The front view of a flower showing the sepals, the petals and the lip with a rounded midlobe fringed along the margin. C. The lateral view of a flower with sepals and petals removed, showing the column, the mentum, and the lip. The basal bracts of the peduncle indicate the development of the inflorescence from an axillary bud. D. The lateral view of a flower with sepals, petals, and half of the lip removed, showing the column, mentum, anther cap, and the stigma. E. The front view of the apical portion of the column showing the anther, the rostellum and the stigma. F. Four pollinia in 2 pairs, without caudicle and viscidium. G. A front-lateral view of the apical portion of the column with the anther cap pushed up, showing the anther with 2 cells on the sides and attached by a filament on the back.

24. **Arundina** Blume 竹葉蘭屬 (FIGURE 29)

Flowers medium-sized, showy, opening one at a time, appearing as if in a terminal raceme; sepals free, subequal, spreading; petals slightly wider than the sepals; lip attached to the base of the column, trumpet-shaped, embracing the column, the apex deeply cleft, disc lamellate, with 5 longitudinal keels; column long, slender, slightly curved and enlarged at the apex; clinandrium elevated, rotundate and 2-dentate; rostellum curtain-like, pointed at the middle of the margin; anther terminal, incumbent, concave; pollinia 8 in 2 groups, discoid, laterally compressed, loosely connected by granular material; stigma semi-orbicular, below the rostellum; ovary cylindric. Capsules cylindric, strongly rostrate. Terrestrial perential orchids, erect. Stems leafy, slightly enlarged at the base. Leaves many, lanceolate, articulate with the sheaths. Inflorescences occasionally branched; bracts small, persistent. One species in Hong Kong, *Arundina chinensis* Blume, growing on grassy slopes or along the banks of streams in exposed areas.

25. **Epidendrum** L. 樹生蘭屬 (FIGURE 30)

Flowers showy, medium-sized, in crowded terminal racemes; sepals and petals similar, free, spreading; lip connate with the column forming a funnel below the stigma, the dilated portion trilobed, the lobes fringed, disc with 2 prominent conic calli; column slender, the apical portion widened and adnate to the base of the lip (or free), the apex with 2 auricles; anther terminal, incumbent, convex, semiglobose, 2-locular, each cell septate; pollinia 4 in 2 pairs, waxy, compressed ovoid, associated with 2 pairs of caudicles consisting of densely imbricate granular delicate scales; viscidium tubular; rostellum bilobed with the viscidium fitting into the space between the lobes; stigma on the roof of the funnel mentioned above; ovary striate-sulcate, glabrous. Epiphytic perennial orchids. Roots long, grayish-white, branched. Stems subterete. Leaves numerous, oblong, thick coriaceous, retuse. Scapes terminal, with many scales; bracts deltoid. Native of tropical America, one species with red-yellow flowers cultivated in Hong Kong, *E. ibaguense* H.B.K.

26. **Nephelaphyllum** Blume 雲葉蘭屬 (FIGURE 31)

Flowers few, non-resupinate, in a loose terminal raceme; sepals and petals subequal, spreading like a fan; lip uppermost, undivided, long and narrow, calcarate, the spur enlarged at the end, the lateral lobes obsolete, midlobe spreading, 2-lobed, erect, disc cristate-lamellate and papillose to the base; column long and stout, winged, the apex erose; clinandrium broad; rostellum fleshy, truncate, recurved; anther terminal, rectangular and 2-cornute; pollinia 8 in 2 groups, waxy, laterally compressed, 2 seriate, the upper ones smaller, loosely associated by glanular basal membrane; stigma discoid; ovary subcylindric-ovoid. Capsules oblong, reflexed, ridged. Terrestrial perennial orchids, with leaves remaining on the third year's growth. Roots few. Rhizomes fleshy, the internodes longer than thick. Leaves ovate, fleshy, produced 1 annually, disarticulation near the truncate base. Scapes emerging from the base of the last internode, with 1 basal and 1 submedian sheaths. One species in Hong Kong, *N. cristatum* Rolfe.

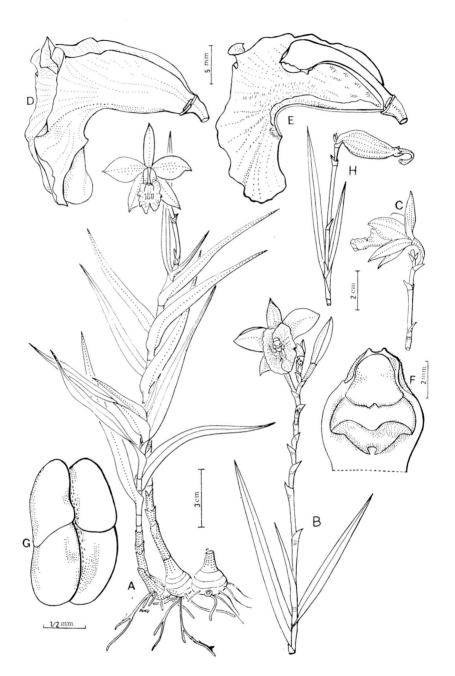

FIGURE 29. *Arundina chinensis*: A. The habit sketch of the species showing a flowering plant attached to an old pseudobulb on the right and a young plant on the left. B. A portion of a flowering plant in a more advanced stage of development, showing many bracts where the flowers have fallen, 1 flower, and a few buds. C. The lateral view of a flower showing sepals, petals, and lip. D. The lateral view of a flower with sepals and petals removed, showing the lip. E. The same with half of the lip removed, showing the column, and 1 keel of the lip. F. The apex of the column showing the anther, the rostellum and the stigma. G. Two pairs of pollinia from half of an anther, the shorter ones being the front pair. H. A portion of the plant showing the fruit with the persistent column.

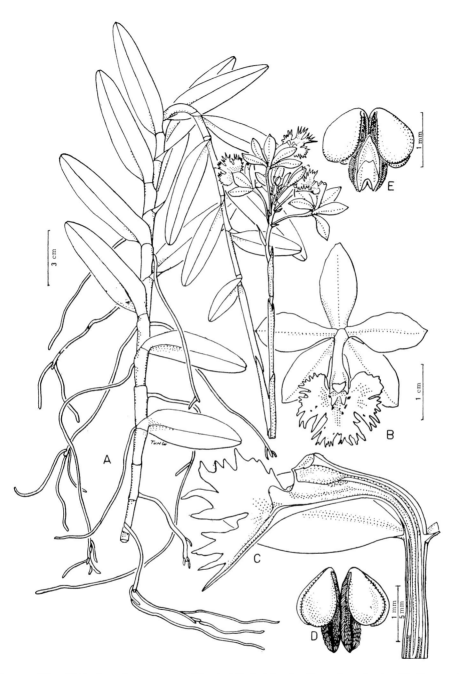

Figure 30. *Epidendrum ibaguense*: A. The habit sketch of a flowering plant with the erect stem bent mechanically to fit into the space on the plate, showing the aerial roots branched only when touching a supporting object, the leaves, and the terminal inflorescence. B. The front view of a flower showing the spreading sepals and petals and the trilobed lip fringed along the margin. C. The longitudinal section of a column and lip showing the union of the lip to the side of the column forming a pouch below the stigmatic cavity. The pouch leads to a papillose tube. D. The front view of the pollinia on soft granular stipes. E. The back view of the pollinia showing a sheath-like viscidium attached to the stipe.

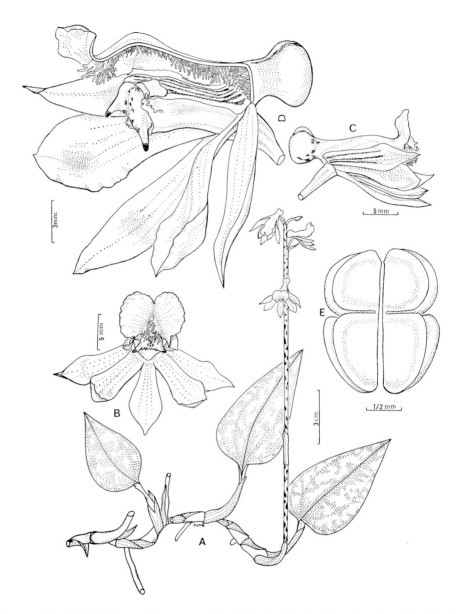

FIGURE 31. *Nephelaphyllum cristatum*: A. The habit sketch of a flowering plant showing the creeping rhizome, representing four years' growth, the position of the flowering shoot, the variegated leaves with very short petioles as indicated by the petiole-like pseudobulb of the fourth year's growth, and the scape with a few non-resupinate flowers. B. The front view of a flower showing the fan-shaped arrangement of 3 sepals and 2 petals, and the superior position of the lip of a non-resupinate flower, with cristate processes in the center of the midlobe. C. The lateral view of a flower showing the inferior position of the odd sepal, 1 lateral sepal, 1 petal, and the calcarate lip curving upward at the tip, depressed at the middle, and with the spur pointing backward and swollen at the apex. D. A flower with half of the lip removed, showing the papillose disc and the cristate processes at the center of midlobe of the lip, the column with an erose collar-like margin, the rectangular anther with 2 horn-like processes pointing downward, the fleshy rostellum curving upward, and the discoid stigma. E. Eight pollinia in four pairs.

27. **Mischobulbum** Schlechter 球柄蘭屬 (FIGURE 32)

Flowers medium-sized, few in a short terminal raceme; sepals and petals subequal, the lateral sepals oblique, connate at the base and adnate to the column-foot forming a mentum; lip fleshy, undivided, subrhomboid, the sides erect, the apex triangular-acute, disc with 3 prominent keels; column long, erect, fleshy, slightly incurved, winged on both sides; rostellum deltoid; anther terminal, ovate, concave, bicornute; pollinia 8 in two groups, each group agglutinated at the base by a sticky mass; stigma transversely concave; ovary subcylindric, striate. Terrestrial perennial orchids with the leaves remaining on the fifth year's growth. Rhizomes short and stout, the internodes thicker than long. Leaves ovate-cordate, produced 1 annually, petiolate, articulation near the cordate base of the lamina. Scapes emerging from one year-old pseudobulbs, with 3 basal and a submedian sheaths. One species in Hong Kong, *M. cordifolium* (Hook. f.) Schlechter.

28. **Tainia** Blume 鄧蘭屬 (FIGURE 33)

Flowers medium-sized, resupinate, in a well-spaced raceme; sepals subequal, spreading, the odd one free, the lateral ones adnate to the short foot of the column forming a mentum; petals narrower than the sepals; lip attached to the apex of the column-foot, slightly gibbous at the base, 3-lobed, the lateral lobes erect, parallel to the column, the apex obtuse, the midlobe suborbicular, entire, disc lamellate; column slender, slightly oblique, dentate; rostellum truncate; anther terminal incumbent, convex, bicornute; pollinia 8 in 2 groups, 2 seriate, waxy, laterally compressed; stigma suborbicular, concave; ovary oblong-obovoid, striate-sulcate. Capsules ellipsoid, rostrate. Terrestrial perennial orchids with slender erect pseudobulbs. Rhizomes short, stout, the internodes thicker than long. Leaves solitary, 1 terminal to each pseudobulb, petiolate, linear-lanceolate. Scape emerging from the side of a pseudobulb, aphyllous, slender and long, with several sheaths at the base; raceme simple. One species in Hong Kong, *T. dunnii* Rolfe.

29. **Ania** Lindl. 安蘭屬 (FIGURE 34)

Flowers medium-sized, in a loose raceme; sepals and petals subequal, spreading; lip 3-lobed, calcarate, attached to the base of the column, lateral lobes erect, obtuse at the apex or triangular-ovate and acute, midlobe ovate or subrotundate, acute; column long, the apical portion winged; clinandrium dentate; rostellum truncate; anther oblong or obovate, convex; pollinia 8 in 2 groups, 2-seriate, the top 2 of each group smaller; stigma suborbicular; ovary cylindric, striate-sulcate. Capsules cylindric. Terrestrial perennial orchids, growing in shade on damp forest floor along streams. Roots many and long. Rhizomes slender; pseudobulbs ovoid, above the ground. Leaves solitary, 1 on the apex to each pseudobulb, petiolate. Scape emerging from the base of a mature pseudobulb before the leaf. Two species in Hong Kong.

Key to the Species

A. Flowers yellow; sidelobes of lip obtuse; anther obovoid, truncate at the base..............
....*A. hongkongensis* (Rolfe) Tang & Wang
AA. Flowers purplish yellow with 3–5 purple lines, with lavender lip dotted purple; sidelobes triangular-ovate and acute; anther oblong, the base obtuse......*A. ruybarrettoi* Hu & Barr.

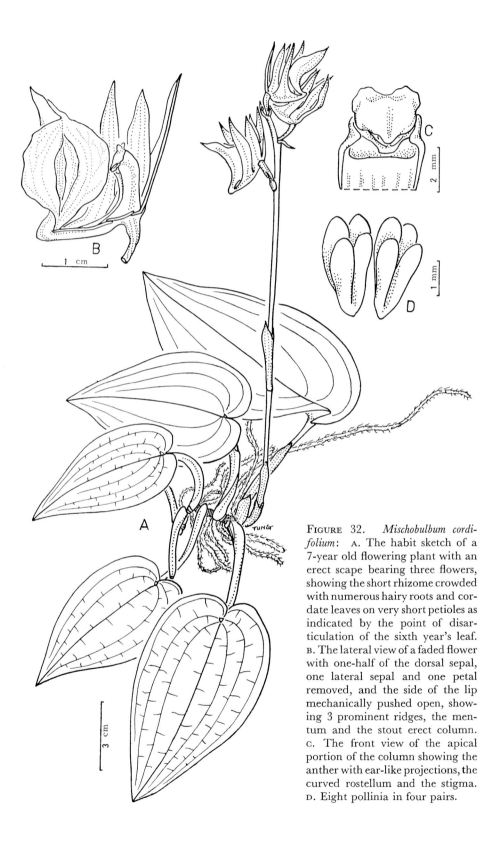

FIGURE 32. *Mischobulbum cordi-folium*: A. The habit sketch of a 7-year old flowering plant with an erect scape bearing three flowers, showing the short rhizome crowded with numerous hairy roots and cordate leaves on very short petioles as indicated by the point of disarticulation of the sixth year's leaf. B. The lateral view of a faded flower with one-half of the dorsal sepal, one lateral sepal and one petal removed, and the side of the lip mechanically pushed open, showing 3 prominent ridges, the mentum and the stout erect column. C. The front view of the apical portion of the column showing the anther with ear-like projections, the curved rostellum and the stigma. D. Eight pollinia in four pairs.

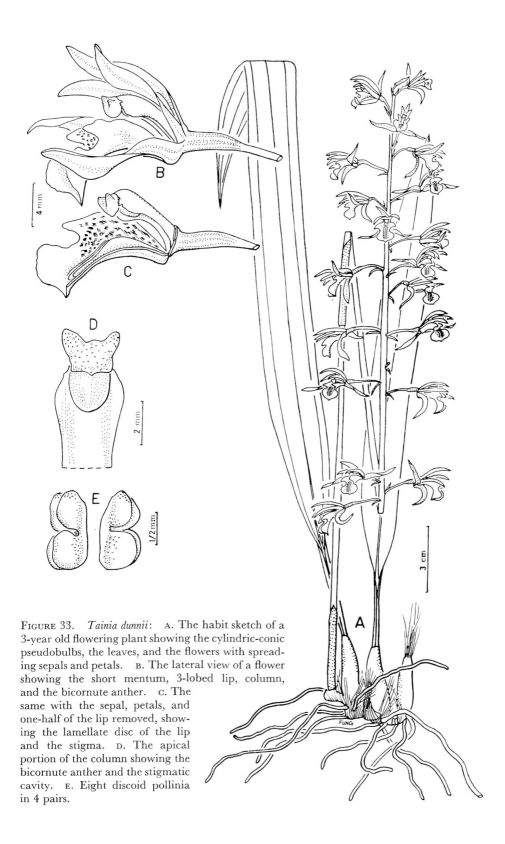

FIGURE 33. *Tainia dunnii*: A. The habit sketch of a 3-year old flowering plant showing the cylindric-conic pseudobulbs, the leaves, and the flowers with spreading sepals and petals. B. The lateral view of a flower showing the short mentum, 3-lobed lip, column, and the bicornute anther. C. The same with the sepal, petals, and one-half of the lip removed, showing the lamellate disc of the lip and the stigma. D. The apical portion of the column showing the bicornute anther and the stigmatic cavity. E. Eight discoid pollinia in 4 pairs.

67

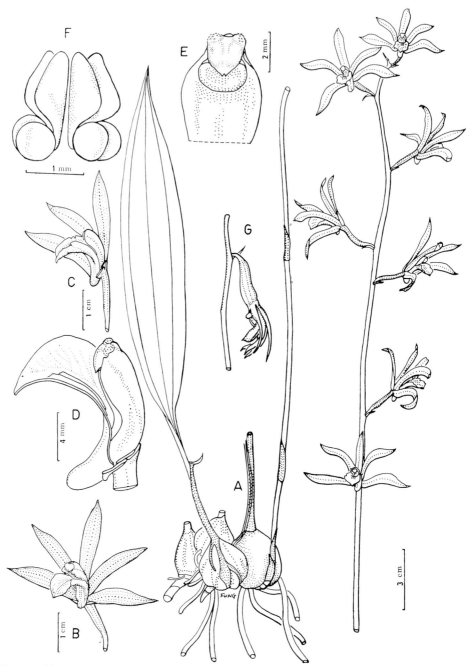

FIGURE 34. *Ania hongkongensis*: A. The habit sketch of a flowering plant with the pseudo-bulbs of four consecutive years, showing the roots, the leaf terminal to one pseudobulb, and the scape basal-lateral to another one. B. The front view of a flower showing the symmetrically arranged sepals and petals, the calcarate lip slightly curved, the column, and the anther. C. The lateral view of a flower with 1 sepal and 1 petal removed, showing the relative positions of the column and the lip. D. The same with all the sepals, petals and one-half of the lip removed, showing the lamellate disc of the lip. E. The front view of an apical portion of the column showing the anther, the stigma, and the broad wing of the column. F. Eight unequal pollinia in 4 pairs. G. A young pendulous fruit.

30. **Pachystoma** Blume 粉口蘭屬 (FIGURE 35)

Flowers medium-sized, resupinate, hairy, in a terminal raceme; sepals subequal, the odd one free, the lateral ones oblique at the base, adnate to the short column-foot forming a mentum; petals slightly narrower than the sepals; lip attached to the column-foot, subgibbous at the base, 3-lobed, lateral lobes erect, rounded, midlobe slightly recurved, retuse, disc fleshy, papillose; column long, arcuate, semiterete and plane in front, enlarged at the apex, forming 2 lateral pouches; clinandrium membranaceous; rostellum obtuse; anther terminal, pilose, 2-lobed; pollinia 8, waxy, obovoid-cuneate, subcoherent at the apex of the slender caudicles; ovary cylindric, hairy. Capsules oblong, ridged and rostrate. Terrestrial perennial orchids, growing on grassy slopes, dormant in winter. Rhizomes flesh, cylindric or yoke-shaped, with 2 buds developing into a leafless scape and a vegetative shoot separately. Leaves linear, chartaceous, plicate-venose. Scape aphyllous, with many brownish cataphylls. One species in Hong Kong, *P. chinense* (Lindl.) Reichb. f.

31. **Eulophia** R. Br. 美冠蘭屬 (FIGURE 36)

Flowers large or medium-sized, in terminal loose racemes; sepals equal, spreading, the odd one free, the lateral ones attached to the side of the column-foot; petals shorter and broader or narrower than the sepals, spreading; lip 3-lobed, attached to the column, slightly contracted at the base and produced into a spur, lateral lobes erect, midlobe broad, almost entire, disc cristate or lamellate; column rather long, with a foot; clinandrium strongly oblique; rostellum subtruncate; anther terminal, incumbent, semi-globose, produced into an obtuse-conic or bilobed appendage; pollinia 4 in 2 pairs, waxy, on a short stipe with viscidium at the base; stigma suborbicular or transversly broad; ovary cylindric. Capsules oblong, pendulous, ridged. Terrestrial perennials growing on grassy hillsides, or saprophytic orchids occurring along the edge of forest floor by streams. Rhizomes lumpy, large, with concentroic rings. Scapes aphyllous, appearing before the leaves, 50–150 cm. tall, with many scales. Three species in Hong Kong, very different in habit and habitat.

Key to the Species

A. Normal leafy plants growing on dry grassy hillside; flowers bright yellow or greenish with pink-white lip.
    B. Flowers yellow; rhizomes subterranean, oblong, 10–12 cm. long, 5 cm. in diameter; leaves elliptic, 20–30 cm. long, 5–10 cm. wide, plicate-venose............*E. flava* Lindl.
    BB. Flowers greenish with white-pink lip; rhizomes partially exposed, green, globose-ovoid, 4–5 cm. in diameter; leaves linear, 15–25 cm. long, 1–2 cm. wide.....*E. sinensis* Rolfe
AA. Saprophytic plants consisting of large subterranean rhizome with concentric black rings and stout flowering scapes; flowers brown-yellow......................*E. yushuiana* S. Y. Hu

32. **Bletilla** Reichb. f. 白及屬 (FIGURE 37)

Flowers medium-sized, showy, in a loose terminal raceme, with deciduous bracts; sepals free, subequal; petals slightly larger, spreading; lip attached to the base of the column, trilobed, lateral lobes erect, slightly embracing the column, midlobe strongly lamellate; column long, without a foot, winged at the apex; clinandrium with broad lateral lobes; rostellum projecting forward forming a roof over the stigma; anther attached to a tooth of the clinandrium, incumbent, ovate;

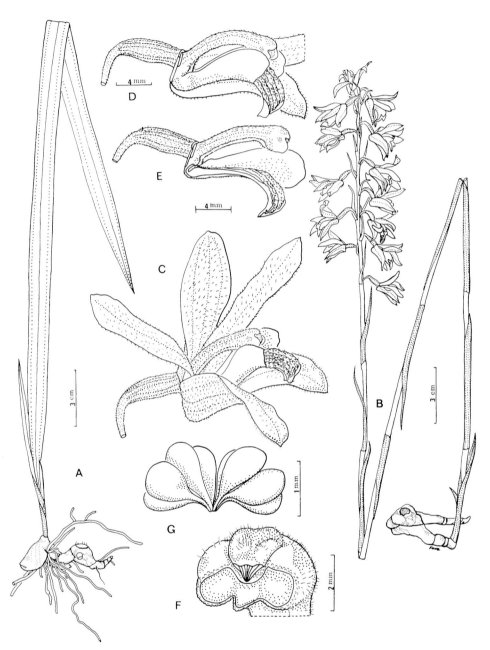

FIGURE 35. *Pachystoma chinense*: A. The habit sketch of a young plant before anthesis showing the rhizome, the root, and the basal leaves. B. A flowering plant showing the yoke-shaped rhizome with two branches, the one in the foreground bearing a terminal bud developed into the flowering shoot, the other one designated to develop into a vegetative shoot. C. The lateral view of a flower showing the pubescent sepals and petals, the column and the hairy ovary. D. The same with the sepals and petals partially removed, showing the column with a short foot, the trilobed lip gibbous at the base, the lateral lobes erect and the midlobe with five papillose longitudinal ridges. E. The same with an additional one-half of the lip removed, showing the fleshy disc papillose on the upper surface, and the erect lateral lobe rounded at the apex. F. The front view of the apical portion of the column showing the anther cap, the rostellum and stigma. G. Eight obovoid pollinia with short caudicles.

pollinia 8, unequal, in two groups of 4 pairs, waxy, oblong and laterally compressed, the base loosely connected by a granular laminiform appendage; viscidium obscure; stigma suborbicular; ovary cylindric, striate-sulcate. Capsules oblong-fusiform, erect, rostrate. Terrestrial perennial orchids with irregularly lobed lumpy rhizomes. Stems of the flowering plants short, leafy, with some cataphylls at the base, gradually changing into leaves, the basal section developing into a rhizome after anthesis. Leaves 2 or 4, appearing with the scape, plicate-venose. Flowers purplish red, white-flowered cultivar rare. One species cultivated for ornamental purposes; escapee observed in Hong Kong Island, *B. striata* (Thunb.) Reichb. f.; rhizome used in Chinese medicine.

33. **Spathoglottis** Blume 苞舌蘭屬 (FIGURE 38)

Flowers medium-sized, in a loose terminal raceme; sepals and petals subequal, free, spreading; lip 3-lobed, attached to the base of the column, slightly saccate at the base, lateral lobes erect, rotundate at the apex, midlobe panduriform, with or without a tooth on each side at the base, the disc with 2 tuberculate-ridged calli near the sinus and a keel running from the calli toward the apex; column long, arcuate, winged near the apex and with or without horn-like appendages; clinandrium short, incurved; rostellum truncate; anthers terminal, incumbent, convex; pollinia 8, subequal, in two groups of 4 pairs, waxy, obovoid; viscidium triangular; ovary pubescent. Capsules oblong, with persistent perianth, prominently ridged. Terrestrial perennial orchids, common and widespread on the arid grassy slopes of Hong Kong, fire-resistant, dormant in winter. Rhizomes compressed globose. Leaves linear, plicate-venose, appearing before the flowers. Scape emerging from a lateral bud of the rhizome, with basal scales; flowers bright yellow or purple; variations evident with the lip and column, population studies needed; 3 species recognized.

Key to the Species

A. Flowers yellow, 2–3 cm. across; flowering bracts 8 mm. or less long, 3 mm. or less wide; lateral sepals 1.5 cm. long (spontaneous in Hong Kong, hillside plants).
    B. Lip 8 mm. long, without teeth at the base of the midlobe; column without horn-like appendages; stigma triangular..*S. fortunei* Lindl.
    BB. Lip 10–13 mm. long, with a small tooth on each side of the base of the midlobe; column with horn-like appendages at the apex; stigma rounded...............*S. pubescens* Lindl.
AA. Flowers purple, 4–5 cm. across; flowering bracts foliaceous, ovate, 2 cm. long, 1 cm. wide; lateral sepals 2–2.5 cm. long (native of Melasia, cultivated).................*S. plicata* Blume

34. **Thelasis** Blume 矮柱蘭屬 (FIGURE 39)

Flowers small, hardly opening, closely arranged in a spike; sepals equal, connivent; petals smaller than the odd sepal; lip undivided, attached to the base of the column, concave; column very short, conic, without wings or foot; rostellum terminal, acuminate; anther subsessile, ovate-acuminate; pollinia 8, in 2 groups of 4, obovoid-globose, attached to an elongated stipe with widened apex and an elliptic viscidium; stigma large, ovate; ovary obconic, very short. Capsules small, oblong. Epiphytic small perennial orchids. Stems short, changing into globose pseudobulbs with 1 or 2 leaves. Scapes aphyllous, with sheathing bracts at the base. One species in Hong Kong, *T. hongkongensis* Rolfe, perhaps the least specialized of all the epiphytic orchids.

35. **Liparis** L. C. Rich. 羊耳蘭屬 (FIGURES 40–41)

Flowers small, numerous, in terminal racemes; sepals free, spreading, the odd one linear, erect, the lateral ones oblique-ovate, often recurved; petals linear or filiform, more or less curved; lip undivided, cuneate at the base, with 2 tuberculate calli, gradually widened and bent at the middle, tongue-like, truncate or dentate, the apex emarginate; column semiterete, elongate, arcuate, the apical portion winged; rostellum truncate; anther terminal, incumbent, subglobose, acute, retuse or acuminate; pollinia 4 in 2 pairs, waxy, obovoid, without obvious viscidium; stigma 1, concave; ovary obconic. Capsules oblong-obovoid. Epiphytic or terrestrial perennial orchids. Stems short, changing to compressed ovoid or cylindric fleshy pseudobulbs. Leaves 1–4, persistent to the second year, ensiform or plicate-venose. Seven species in Hong Kong.

Key to the Species

A. Pseudobulbs cylindric, 9–12 cm. long; leaves elliptic (except *L. longipes*).
  B. Leaves linear, 1.5 cm. wide; lip rotundate at the apex; pseudobulbs 8–9 cm. long, 5–7 mm. across the middle, the longest internodes 6–8 cm. long......*L. longipes* Lindl.
  BB. Leaves elliptic, 2.5–6.5 cm. wide; lip dentate or notched at the apex; pseudobulbs 4.5–12 cm. long, 1–1.5 cm. in diameter, the longest internodes 2.5–6 cm. long.
    C. Flowers fresh liver-color; apical portion of lip 15 mm. across, the margin irregularly dentate-fimbriate...*L. macrantha* Rolfe
    CC. Flowers yellowish green or lightly tinged purple; apical portion of lip 4.5–7 mm. across, shallowly undulate-dentate..........................*L. nervosa* (Thunb.) Lindl.
AA. Pseudobulbs ovoid or subglobose, 1–2.5 cm. long; leaves linear.
  B. Leaves solitary, 1 to each pseudobulb; apical wings of the column broad and auriculate at the base.
    C. Anther retuse at the apex; odd sepal 7 mm. long...............*L. plicata* Fr. & Sav.
    CC. Anther long produced at the apex; odd sepal 11 mm. long..........................
    ....*L. ruybarrettoi* Hu & Barr.
  BB. Leaves 2–4 to each pseudobulb; apical wings of column narrow, not auriculate at the base.
    C. Lateral sepals spreading, 10–12 mm. between the apexes; lip greenish yellow, the apical portion fan-shaped; calli roundish; flowering in January–February.
    ....*L. chloroxantha* Hance
    CC. Lateral sepals twisted forward, below the lip, the apices close or overlapping; lip tinged rose-purple to maroon, the apical portion suborbicular and truncate; calli conic, flowering in June.......................................*L. odorata* (Willd.) Lindl.

FIGURE 36. *Eulophia sinensis*: A. The habit sketch of the vegetative growth showing a new shoot emerging from the base of a pseudobulb that bore 2 flower scapes earlier in the growing season. B. The habit sketch of a flowering plant which has no leaves during anthesis. C. The habit sketch of a large pseudobulb after anthesis showing the relative positions of 4 scapes, and the new leaves. D. The front view of a flower with the anther cap removed, showing the sepals, the petals, the lip with small lateral lobes and a large midlobe crisped along the margin and with elongated filiform tubercles and processes in the middle, and the apex of the column with 2 pollinia exposed. E. The lateral view of a flower with 1 sepal and 1 petal removed, showing the relative position of the calcarate lip. F. The same with 1 sepal and 1 petal removed and the lip spread out, showing the longitudinal ridges and the papillose processes. G. The apical portion of the column showing the anther with the ear-like apical papillose lobes, the pollinia on a stout stipe, the rostellum, and the stigma. H. The apical portion of the column with the anther cap removed, showing the triangular clinandrium with sharply pointed apex, and 2 pollinia on a stout stipe. I. A fruiting branch with 4 pendulous capsules, the lowermost one commences to open.

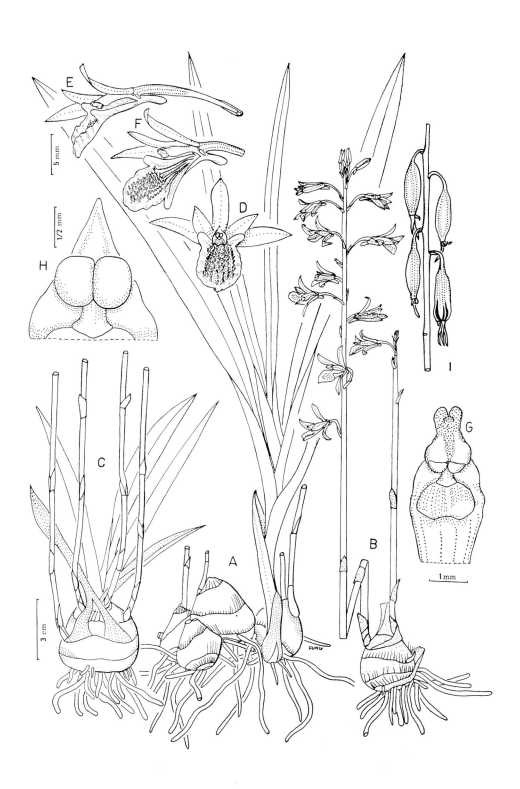

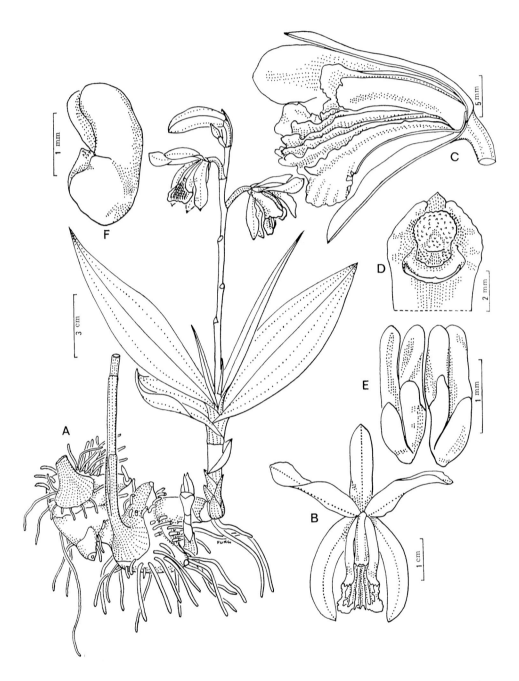

FIGURE 37. *Bletilla striata*: A. The habit sketch of old rhizomes and new shoots of a flowering plant showing the irregularly lobed rhizomes and a leafy shoot terminated by scape. B. The front view of a flower showing the sepals, the petals, the lip with lamellate disc, and the erect lateral lobes partially covering the column. C. The lateral view of a flower with one-half of the dorsal sepal, a petal, the lateral sepals and 1 lateral lobe of the lip removed, showing the lamellate disc of the lip and the column broadly winged at the apex. D. Apical portion of the column showing the anther, the rostellum and the stigma. E. The dorsal view of 8 unequal pollinia in 2 groups of 2 pairs. F. The lateral view of 1 group of the pollinia.

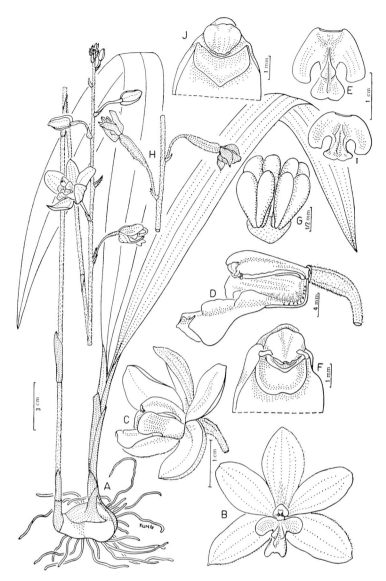

FIGURE 38. *Spathoglottis*: A–H. *Spathoglottis pubescens*: A. The habit sketch of a flowering plant showing the vegetative growth and the position of the scape. B. The front view of a flower showing the spreading sepals and petals, the trilobed lip, and the column. C. The lateral view of a flower showing the sepals, petals, the relative position of the column, and the lip slightly saccate at the base. D. The same with the sepals, petals, and a lateral lobe of the lip removed, showing the calli and ridge at the base of the midlobe. Note the hairs on the ridge. E. A lip as seen from the top, 10–13 mm. long, showing the lateral lobes and the midlobe, with the calli at the narrowed section of midlobe. F. A front view of the apical portion of the column showing the appendages on the apex of the column, the anther, the triangular rostellum, and the rounded stigma. G. Eight pollinia in 4 pairs loosely associated by a granular membrane at the end. H. The habit sketch of a fruiting branch showing the young capsules with persistent perianth. I–J. *Spathoglottis fortunei*: I. Lip, as seen from the top, 8 mm. long, showing the lateral lobes and midlobes, with the calli at the center of the midlobe. J. A front view of column showing the lack of apical appendage, the anther, the rostellum, and the triangular stigma.

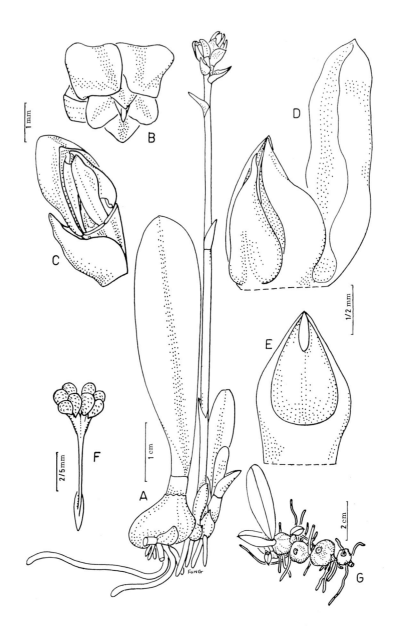

FIGURE 39. *Thelasis hongkongensis*: A. The habit sketch of a flowering plant showing the roots, a mature leaf on a pseudobulb, and the relative positions of a scape with flowers and the new shoot. B. The front view of the non-resupinate flower showing the odd sepal being lower-most and lateral sepals at the top, the petals, the lip, and the apex of the column. C. The lateral view of a flower with the odd sepal and 1 lateral sepal removed, showing the bract, 1 thick lateral sepal, the petals, and the lip. D. The lateral view of the short column showing the undivided lip, the anther, and the beak-like rostellum. E. Apical view of the column showing the stigmatic cavity and the viscidium. F. Eight pollinia in 4 pairs, on elongated stipe, with viscidium. G. Habit sketch of a 4-year old vegetative shoot showing the current year's pseudobulb bearing 2 leaves and the older ones without a leaf.

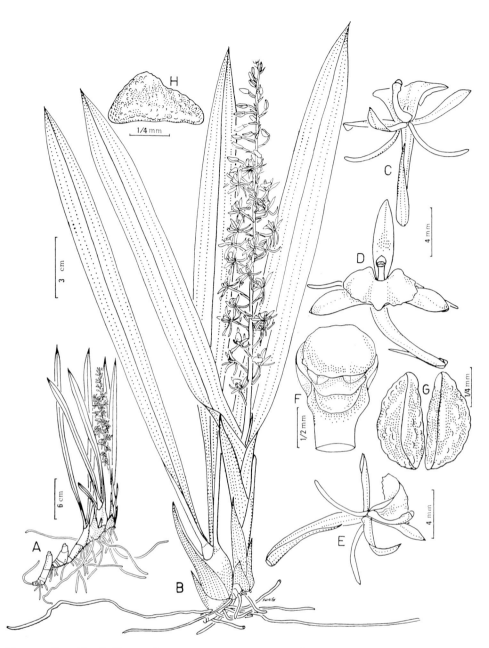

FIGURE 40. *Liparis chloroxantha*: A. The habit sketch of a flowering plant showing the growth of 5 consecutive years, the 2 older pseudobulbs without leaves, the third and second years' growth with leaves remaining on the pseudobulbs, and the current year's plant with a flowering scape and the pseudobulb not yet evident. B. The habit sketch of a two-year old plant showing a mature pseudobulb, a flowering shoot with imbricate cataphylls, 2 fully grown leaves, and a flowering scape with the pseudobulb not as yet fully developed. C. The abaxial view of a flower showing the non-resupinate position with the lip in a superior position, the 3 sepals, and the narrow petals. D. The face view of a flower showing the broad apical portion of the lip. E. The top view of flower showing the relative position of the sepals, petals, lip, and the column. F. The apical portion of the column showing the anther cap with a portion of the pollinia, the rostellum, and the stigma. G. Two pairs of the pollinia slightly pushed apart, showing the uneven surface. H. The side view of 1 pollinium showing the shape and the uneven surface.

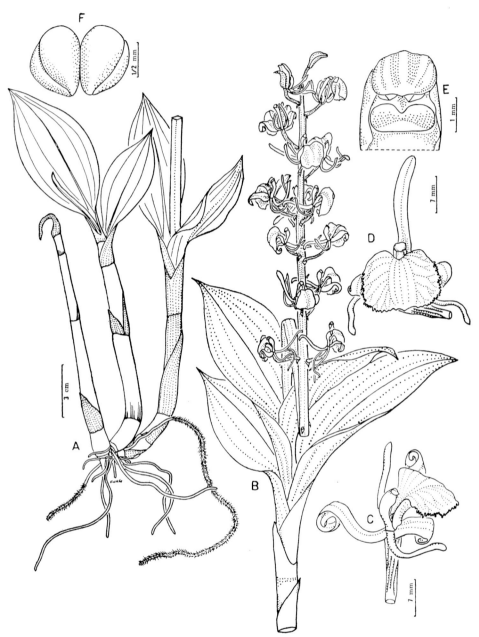

FIGURE 41. *Liparis macrantha*: A. The habit sketch of a plant showing the cord-like roots and three years' growth with the third year's pseudobulb leafless, the second year's stem with leaves, and the current year's stem bearing a terminal inflorescence. B. The current year's growth of a large plant with 4 leaves, and a scape with 3 bracts, showing the non-resupinate and some partly resupinate flowers. C. The lateral view of a flower showing the undivided tongue-like lip with 2 calli one on each side at the base, the arcuate and winged column, the sepals curved at the apex, the linear petals, and the ridged and winged ovary. D. The front view of a flower showing the broad apical portion of the lip irregularly dentate-fimbriate along the margin, the sepals, the petals, and the apex of the column. E. The apex of the column showing the subglobular anther cap, a portion of the pollinia, the truncate rostellum, and the cup-shaped stigma. F. Four pollinia in 2 pairs, without viscidium.

36. **Malaxis** Swartz 沼蘭屬 (Figure 42)

Flowers very small and numerous, non-resupinate, in a terminal raceme; sepals free, spreading and often curved; petals filiform, curved forward; lip sessile, erect, spreading or the sides erect, auriculate at the base, with a depression near the column, the apex acute or bilobed; column very short, sunk in the sinus of of the depression of the lip, terete, truncate at the apex; rostellum truncate; anther suborbicular, sessile, erect; pollinia 4 in 2 pairs, waxy, oblong-obovoid, without caudicle; viscidium a narrow mass lining the truncate rostellum; stigma concave; ovary obconic, striate-sulcate. Capsules ellipsoid, the apex truncate, with the column surrounded by persistent remains of the sepals and petals. Terrestrial perennial orchids growing in rocky wet areas along the streams; dormant during the cool dry season. Stems leafy, changing into elongated subconic or cylindric pseudobulbs. Leaves large, plicate-venose, chartaceous. Four species in Hong Kong.

Key to the Species

A. Lip shield-like, without erect sides, the apex bilobed; column short, the length and the width subequal.
  B. Odd sepals linear; apex of lip rotundate; margin of leaves wavy..........................
  ....*M. allanii* Hu & Barr.
  BB. Odd sepals oblong; apex of lip acuminate; margin of leaves smooth....................
  ....*M. acuminata* D. Don var. *biloba* (Lindl.) Hunt
AA. Lip erect on both sides, the apex triangular, entire and acute; column terete, the length twice as long and wide.
  B. Pseudobulb subglobose-conic, 2 cm. long; odd sepal linear, 1 mm. across...............
  ....*M. parvissima* Hu & Barr.
  BB. Pseudobulb cylindric, 12–13 cm. long; odd sepal oblong, 5 mm. across.................
  ....*M. latifolia* J. E. Sm.

37. **Cattleya** Lindl. 布袋蘭屬 (Figure 43)

Flowers showy, solitary or few in a small cluster; sepals subequal, free, spreading; petals broader than the sepals; lip trumpet-shaped, attached to the base of the column and concealing the latter by the curved basal portion, the lateral lobes obscure, the midlobe expanded, with crisped margin and bilobed apex; column elongated and incurved, slightly enlarged at the apical portion; clinandrium oblique, dentate; anther terminal, incumbent, subglobose, 2-locular, the cells longitudinally septate; pollinia 4 in 2 pairs, waxy, strongly compressed, parallel, each pair loosely attached to a thin sheet of granular appendage; rostellum truncate; stigma obovate, concave; ovary cylindric, striate-sulcate. Capsules oblong-ovoid, with prominent ridges, rostrate. Epiphytic orchids. Rhizomes short, covered by persistent cataphylls; pseudobulb oblong. Leaves solitary, or 2, apical to the pseudobulb, thick coriaceous, band-like, rotundate or retuse at the apex. Many cultivars introduced for ornamental purposes.

Key to the Species Examined

A. Flowers delicate purplish-rose suffused white; sepals oblong; petals elliptic, lip crisped and emarginate, amethyst purple.....................................*C. lueddemanniana* Reichb. f.
AA. Flowers white or tinted mauve; sepals oblong-lanceolate; petals oblique ovate, crisped; lip spreading, much crisped and indented, rich crimson-purple...........*C. mendelii* Backh.

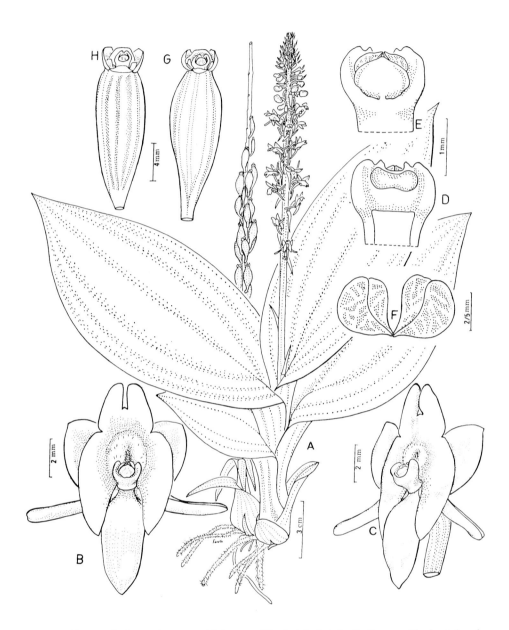

FIGURE 42. *Malaxis acuminata* var. *biloba*: A. The habit sketch of a 2-year old plant showing a fruiting shoot with persistent cataphylls, and a flowering shoot with mature leaves. B. The front view of a flower showing the non-resupinate position with the odd sepal pointing downwards, the broad lateral sepals, the narrow petals, the shield-like lip bilobed at the apex and depressed at the middle, the short column with apical arms, the anther cap, and the pollinia. C. The lateral view of a flower showing the relative position of sepals, petals, lip, broad column, and the anther. D. The front view of the column showing its juncture with lip, the stigma, and the rostellum. E. The back view of the column showing the apical arms, the anther cap, and the pollinia. F. Four pollinia in two pairs, without viscidium. G. A fruit showing the persistent portions of the sepals, petals, lip, and the broad column with open arms (*Hu 12449*). H. Another fruit showing the persistent portions of the sepals, petals, lip, and the broad column with the arms close to the central axis (*Hu 12451*).

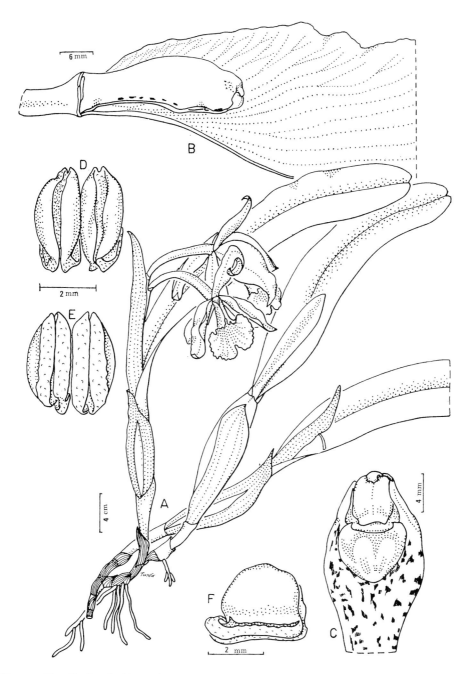

FIGURE 43. *Cattleya lueddemanniana*: A. The habit sketch of a flowering plant showing the branched rhizome with roots, the large and small leaves terminal to the oblong-ellipsoid pseudobulbs, and the position of the flower cluster with 2 flowers. B. The lateral view of a flower with all the sepals and petals, and a large portion of the lip removed, showing the column with a terminal anther. C. The apical portion of the column showing the anther, rostellum and stigma. D. The top view of two pairs of strongly compressed pollinia loosely attached at the base by a granular thin sheet. E. The bottom view of the same. F. The lateral view of one pair of pollinia with the basal granular appendage.

38. **Pholidota** Lindl. 山桃蘭屬 (FIGURE 44)

Flowers rather small, non-resupinate, in a terminal raceme; sepals subequal, ovate, spreading, slightly concave; petals slightly narrower than the sepals; lip strongly concave, subsaccate at the base, the lateral lobes small and erect, midlobe straight or recurved; column stout, the apex broadly winged; rostellum truncate, projected forward covering the stigmatic cavity; anther incumbent, subglobose; pollinia 4, waxy, ovoid, and acuminate; viscidium apical; stigma discoid; ovary cylindric, striate. Capsules elliptic-obovoid, strongly ridged and grooved. Epiphytic perennial orchids growing on damp rocks or rarely on trunks of trees along streams. Rhizomes covered by imbricate scales. New shoots emerging from the base of the pseudobulbs, each bearing a terminal raceme, after anthesis a joint behind the inflorescence changing into a pseudobulb with 2 leaves. Leaves terminal to pseudobulb, evergreen, elliptic or lanceolate. Flowering shoots horizontal, bearing green cataphylls and two underdeveloped leaves at anthesis, becoming erect and with the leaves and pseudobulbs developed after flowering. Two species in Hong Kong.

Key to the Species

A. Leaves elliptic, 2–4 cm. wide; racemes hanging; flowers 2 cm. across; sepals ovate; midlobe of the lip recurved...*P. chinensis* Lindl.

AA. Leaves linear-lanceolate, 5–8 mm. wide; racemes erect; flowers 5 mm. across; sepals ovate-oblong; midlobe of the lip short, acute, not recurved................*P. cantonensis* Rolfe

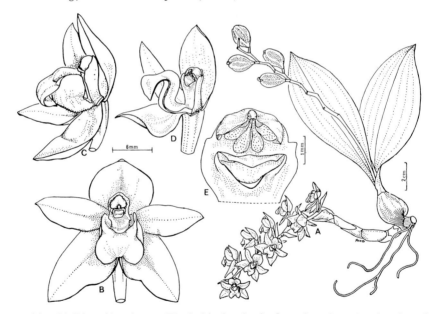

FIGURE 44. *Pholidota chinensis*: A. The habit sketch of a flowering plant showing the orientation of the scape from the base of a 2-year old pseudobulb with the last year's scape and 4 fruits remaining between the leaves. The pendant position of the raceme and the position of the flowering bract indicate the non-resupinate condition of the flower. B. The front view of a flower showing the sepals, petals, lip, column, anther, rostellum, and the stigma. C. The lateral view of a flower with a petal removed, showing the relative positions of the column and the saccate lip. D. The same with a sepal, a petal, and one-half of the lip removed, showing the column and the attachment of lip. E. The apical portion of the column with the anther slightly tipped back, showing the 4 pollinia, and the rostellum curving over the stigmatic cavity.

39. **Coelogyne** Lindl. 貝母蘭屬 (FIGURE 45)

Flowers medium-sized, opening one at a time, 2 or 3 in a terminal raceme; sepals free, subequal, spreading; petals narrower than or rarely similar to the sepals; lip attached to the base of the column, 3-lobed or undivided, the base concave and slightly saccate, lateral lobes obsolete or erect, partially embracing the column; midlobe expanded and fimbriate or entire and acute, the disc lamellate; column elongated, winged at the apical portion, without foot; clinandrium oblique, membranous, dentate; rostellum membranous, projecting forward and arching over the stigma; anther attached under the margin of the clinandrium, incumbent; pollinia 4 in 2 pairs, waxy, obovoid, connected at the apex by a granular thin sheet; stigma suborbicular; ovary cylindric. Capsules ellipsoid, ridged, rostrate. Ephiphytic perennial orchids creeping on the surface of boulders propped and anchored by slender tough roots. Leaves 2, apical to the oblong-ellipsoid pseudobulb, coriaceous. Scape between leaves, with several persistent basal scales; flowering bracts deciduous. Two species in Hong Kong.

FIGURE 45. *Coelogyne fimbriata*: A. The habit sketch of a flowering plant showing the branching system of the rhizome with persistent scales, the pseudobulbs, the leaves, the flowers, and fruit. B. The front view of a flower showing the ovate-elliptic sepals, the linear petals, and the relative positions of the lip and column. C. The lateral view of a flower showing the sepals, the petals, the slightly saccate lip with erect lateral lobes and broad fimbriate midlobe with 2 cristate ridges. D. The same with sepals and petals removed, showing the lip and the column. E. The same with one-half of the lip removed, showing one longitudinal ridge of the midlobe, the winged column, the anther, and the stigma. F. The front view of an apical portion of the column showing the irregularly dentate apex, the lateral wings, the anther cap, the eaves-like rostellum, and the stigma. G. Four pollinia in 2 pairs attached to a granular sheath.

Key to the Species

A. Petals linear, narrower than the sepals; lip trilobed and fringed, the midlobe expanded, subtruncate at the apex...*C. fimbriata* Lindl.

AA. Petals lanceolate, similar to the sepals; lip entire, lanceolate and acute at the apex......
....*C. leungiana* S. Y. Hu

40. **Cirrhopetalum** Lindl. 捲瓣蘭屬 (FIGURE 46)

Flowers resupinate, of moderate size, graceful and attractive, in a simple umbel arranged in form of a fan; sepals unequal, the odd one free, ovate or ovate-oblong, fornicate, the lateral ones 2–6 times longer than the odd one, adnate to the column foot and forming a mentum at the base, then twisted, with the upper margins meeting before the lip, forming an coherent platform; petals similar to and smaller than the odd sepal, lip tongue-like, fleshy, mobile, contracted at the base and hinged to the end of the column-foot, incumbent, recurved; column short, stout, the base produced into an elongated and curved foot, the sides winged, and the wings extended into 2 prominent arms at the apex; clinandrium truncate or dentate; rostellum truncate; anther depressed, hemispherical, deciduous; pollinia 4 in 2 pairs, waxy, the inner one of each pair smaller, without caudicles or viscidium; stigma strongly concave, with a large glossy sticky gland on the upper margin below the rostellum and two small ones on the opposite margin; ovary obovoid-cylindric. Capsules oblong-ellipsoid, rostrate, ridged. Epiphytic perennial orchids. Rhizomes slender. Pseudobulbs ovoid, slightly flattened on one side. Leaves solitary, terminal to the pseudobulb, fleshy, lasting for three years. Scape emerging from the base of the flattened side of a pseudobulb, with basal and submedian scales. Five species in Hong Kong, growing on rocks or trunks of trees along streams.

Key to the Species

A. Lateral sepals less than 3 times the length of the odd one; petals without cilia or seta.

    B. Flowering scape emerging from one-year old pseudobulb with all the cataphylls deteriorated and disappeared; sepals, petals, and lip with red longitudinal stripes on yellow background; anther cap smooth..........................*C. tigridum* (Hance) Rolfe

FIGURE 46. *Cirrhopetalum tseanum*: A. The habit sketch of a flowering plant showing three years' growth with numerous roots below the pseudobulbs. Notice the flowering shoot emerging from the base of the second year's pseudobulb with imbricate scales at the base, and 2 sheathing sterile bracts at the lower portion. The pseudobulb of current year's growth is surrounded by imbricate membranous cataphylls. B. The lateral view of a flower showing the cucullate dorsal sepal, the lateral sepals forming a mentum at the base and then twisted upward with their upper margins meeting in front of the lip, the oblique petal, a portion of the column with anther, a portion of the lip, and the ovary. C. A portion of a flower seen from above, showing the fringed odd sepal caudate and filiform at the apex, the petals, the basal portion of the twisted lateral sepals, and the lip. D. The ventral view of a flower showing the mentum, the twisted lateral sepals recurved at the apex, and the tip of the lip. E. The front view of the column and lip showing the wings, the anther, the fleshy recurved lip with a median-longitudinal ridge and a groove on each side. F. The lateral view of the same, showing the elongated and curved column foot and the attachment of the movable fleshy lip which has a median-longitudinal ridge and 2 grooves. G. The apical portion of the column viewed from the front and slightly tipped back, showing the anther, the rostellum, and the stigmatic cavity with 3 marginal glands, a large one above and situated below the rostellum, and 2 smaller ones on the angles of the opposite margin. H. Four pollinia in 2 obovoid pairs, the outer one of each pair larger.

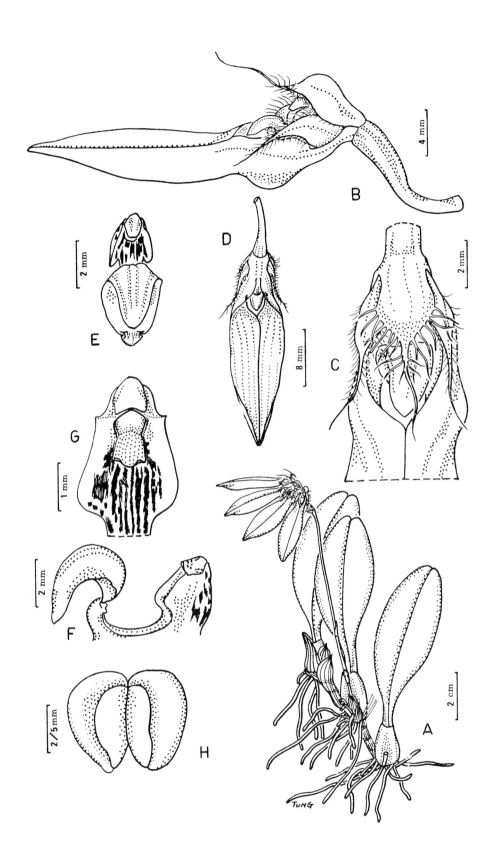

BB. Flowering scape emerging from the base of newly formed pseudobulb covered by imbricate cataphylls; sepals, petals, and lip uniform in color...*C. bicolor* (Lindl.) Rolfe

AA. Lateral sepals 5–6 times longer than the odd sepal; petals ciliate or each with a seta at the apex.

    B. Leaves 4 cm. or more wide; petals oblong-falcate, each with a seta at the apex; scape emerging from the base of newly formed pseudobulb covered by cataphylls; column with elongated bicuspidate arms...............................*C. delitescens* (Hance) Rolfe

    BB. Leaves 3 cm. or less wide; petals ciliate; scape emerging from the base of pseudobulb with cataphylls all deteriorated and disappeared; column arms short and acute.

        C. Flowers salmon pink; lateral sepals 5 cm. long, with both the upper and the lower margins coherent beyond the mentum, forming a tube about 1.5 cm. long; cilia of the odd sepal and petals clavate.....................................*C. miniatum* Rolfe

        CC. Flowers yellow maculate purplish red; lateral sepals 2 cm. long, with the lower margin free beyond the mentum, the upper margin coherent forming a platform; cilia of the odd sepal and petals long and wavy, setaceous............................
....*C. tseanum* Hu & Barr.

41. **Bulbophyllum** Thouars 石豆蘭屬 (FIGURE 47)

Flowers small, in a subumbelliform raceme, or medium-sized and solitary; sepals subequal, spreading, the odd one free, the lateral ones obliquely dilated at the base and adnate to the column-foot forming a mentum; petals smaller than the sepals; lip attached to the column-foot, incumbent, mobile, fleshy, tongue-like, recurved, 2-lamellate, often warty; column short, stout, produced into a foot at the base, winged at the apical portion and extended into 2 arms; clinandrium short, 2 dentate; rostellum truncate; anther terminal, incumbent, depressed, hemispherical or obtuse-conic, deciduous; pollinia 4 in 2 pairs, unequal, the inner one of each pair smaller, waxy, without caudicles or viscidium; stigma concave, occupying the entire front of the column; ovary obconic. Capsules oblong. Epiphytic perennial orchids. Rhizomes slender, with 3–4 internodes, rooted below the pseudobulb. Pseudobulbs cylindric, oblong or ovoid, often compressed or slightly flattened on one side near the base. Leaves solitary, apical to a pseudobulb, coriaceous, oblong, rarely ovate-oblong, rounded or retuse at the apex. Scape filiform, emerging from the base at the flat side of the pseudobulb, with 2–4 basal bracts and some median and supermedian bracts; pedicels filiform. Five species spontaneous in Hong Kong.

Key to the Species

A. Flowers solitary, rarely 2; pseudobulbs cylindric; odd sepals broad-ovate, acute; lip strongly recurved...*B. ambrosia* (Hance) Schlechter

AA. Flowers in subumbelliform clusters; pseudobulbs ovoid or oblong; odd sepals lanceolate or ovate and caudate; lip tongue-like, moderately recurved.

    B. Pseudobulbs crowded together; sepals broadly ovate and caudate; petals dentate along the margin; anther conic, ridged................................*B. levinei* Schlechter

    BB. Pseudobulbs separated by slender rhizomes; sepals lanceolate; petals entire; anther ovoid.

        C. Umbel large, containing 4–15 flowers; pseudobulb oblong; lip papillose.........
....*B. odoratissimum* Lindl.

        CC. Umbel small, with 3 or 4 flowers; pseudobulb ovoid; lip not papillose.

            D. Petals linear; flowers loosely arranged on a raceme with distinctive rachis; flowering bracts over one-half of the length of the pedicels......................
....*B. youngsayeanum* Hu & Barr.

            DD. Petals ovate; flowers in an umbelliform cluster; flowering bracts one-third or less as long as the pedicels.....................................*B. radiatum* Lindl.

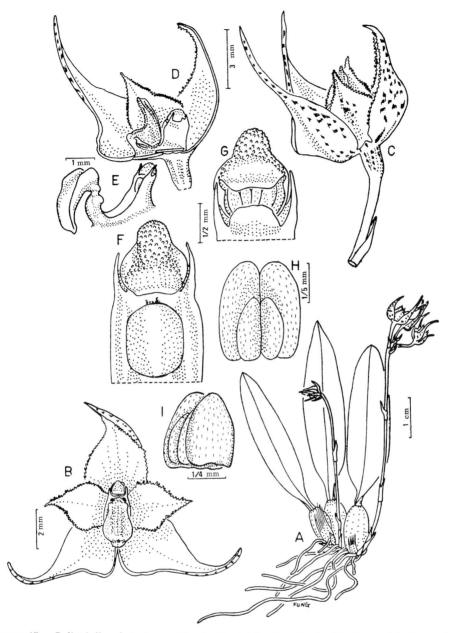

FIGURE 47. *Bulbophyllum levinei*: A. The habit sketch of a plant with 3 pseudobulbs each bearing a single leaf, showing the roots and the flowering scapes. B. The front view of a flower showing the ciliate petals and odd sepal, the entire lateral sepals, the lip, and the column. C. The lateral view of a flower showing the sepals, petals, a portion of the lip, the short ovary, and bract. D. The same with half of the dorsal sepal, 1 lateral sepal, and 1 petal removed, showing the short mentum, the movable thick and papillose lip, and the column. E. The front-lateral view of the column showing the foot, the movable tongue-like lip, and the anther. F. The front view of the column showing the arms, anther, rostellum and stigma. G. The apex of the column with the anther cap slightly pushed up, showing the ventral position of the pollinia, the rostellum, and the horn-like arms. H. Four unequal pollinia in 2 pairs. I. The same in lateral view.

42. **Eria** Lindl. 毛蘭屬 (FIGURE 48)

Flowers medium-sized or small, non-resupinate, in a terminal raceme, rarely solitary (due to reduction); sepals subequal in length, the odd one symmetrical, concave, the lateral sepals oblique, adnate to the foot of the column forming a mentum; petals similar to or narrower than the sepals; lip 3-lobed, attached to the end of the column-foot, recurved, the sides erect, the midlobe lamellate; column short, broad, winged, the base produced into a straight or curved foot; clinandrium truncate or sinuate; rostellum obtuse; anther terminal, incumbent, concave; pollinia 8 in 2 groups, waxy, obovoid, pyriform, the apex coherent slightly by a viscid mass; stigma suborbicular, or the lower margin 3-lobed; ovary subcylindric or obovoid. Capsules oblong, ridged. Epiphytes growing on boulders or rarely on loose soil at the bases of damp rocks. Roots many and long. Pseudobulbs crowded. Leaves 1–4 to each pseudobulb, coriaceous or chartaceous. Four species in Hong Kong.

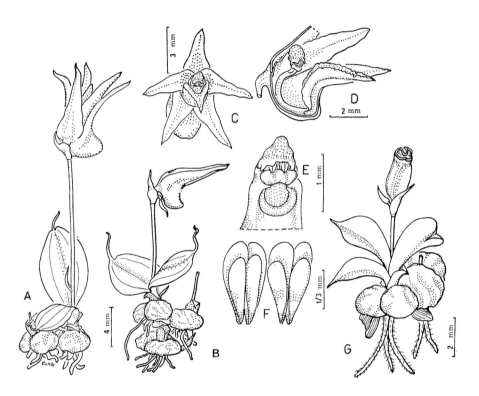

FIGURE 48. *Eria sinica*: A. The habit sketch of plants of three years' growth showing the dorsiventrally compressed pseudobulbs, and the current year's growth with 2 leaves terminated by an apical scape with 1 flower. B. The habit sketch of a colony showing the pseudobulbs with rhizome, and a scape with 2 buds (1 immature), the mature one showing the large saccate base of the lateral sepals associated with the mentum. C. The front view of a flower showing the sepals, petals, lip, column, anther, and the prominent sac at the base of the lateral sepals. D. The lateral view of a flower with half of the dorsal sepal, 1 lateral sepal, and 1 petal removed, showing the short column, the mentum, and the movable tongue-like lip. E. The front view of the column showing the anther, pollinia, a large, smooth and recurved rostellum, and stigmatic cavity with a triangular lobe above and 2 recurved lobes at the lower margin. F. Eight pollinia in 2 groups G. The habit sketch of a colony with a fruit, showing the paired pseudobulbs formed at the base of the leafy shoot.

Key to the Species

A. Plants very small, less than 1 cm. high; pseudobulbs compressed-globular, 2–3 mm. in diameter, always appearing in pairs; flowers solitary; scape filiform; perianth 2–3 mm. long...*E. sinica* Lindl.

AA. Plants 10–20 cm. high; pseudobulbs oblong-ovoid, 1–3 cm. in diameter; flowers several to many, in loose racemes; scape terete; perianth 10 mm. or more long.

    B. Plants growing on soil along damp banks of streams under trees; leaves chartaceous, with several prominent veins; flowers many, 10 or more to each raceme; bracts deciduous...*E. corneri* Reichb. f.

    BB. Plants growing on the surface of exposed boulders in arid places; leaves leathery and fleshy, with visible midrib only; flowers 3–5 to a raceme; flowering bracts persistent.

        C. Flowering bracts exceeding the pedicels and the flowers; leaves solitary, 1 to each pseudobulb, 3 cm. or more wide; sepals glabrous...............*E. rosea* Lindl.

        CC. Flowering bracts shorter than the pedicels; leaves 2–4 to each pseudobulb, 1.5–2.5 cm. wide; sepals white woolly..*E. flava* Lindl.

43. **Sophronitis** Lindl. 貞蘭屬 (Figure 49)

Flowers showy, solitary, terminal to a scape; sepals oblong, subequal, the lateral ones proximate and near the lip; petals broader than the sepals, elliptic, spreading; lip undivided, the sides erect, connivent and covering the column, the apex triangular, acute; column short, subterete, the apical portion petaloid and winged, the apex armed; clinandrium small, with 2 teeth; rostellum truncate; anther terminal, incumbent, convex; pollinia 8, waxy, 2-seriate, parallelly compressed; stigma suborbicular; ovary slender, 3–4 cm. long. Epiphytic perennial orchids. Rhizomes stout, 4–5 joints, the last internode changing into an ellipsoid pseudobulb after anthesis. Cataphylls sheathing. Leaves solitary, 1 to each pseudobulb, ovate-oblong,

FIGURE 49. *Sophronitis grandiflora*: The habit sketch of a flowering plant showing the root, rhizome, the old and new pseudobulbs, the variation in the shapes and sizes of the leaves, the position of the scape and the solitary flower with undivided lip ($\times \frac{2}{3}$).

89

coriaceous. Flowers bright red. One species introduced from Brazil, *S. grandiflora* Lindl.

44. **Acanthephippium** Blume 鎧覃花蘭屬 (FIGURE 50)

Flowers medium-sized, hardly opening, in a loose raceme axillary to a cataphyll of a developing new shoot; sepals fleshy, coherent and forming an oblique ventricose-urceolate tube, the odd sepal linear, the lateral ones oblique, adnate to the column-foot forming a mentum, reflexed at the apex; petals largely concealed, narrower than the lateral sepals; lip short and broad, attached to the elongated and curved column-foot, movable, 3-lobed, the lateral lobes rectangular, erect, midlobe short, recurved, disc lamellate; column thick, winged, the base produced to an incurved foot; clinandrium short, membranaceous; rostellum truncate; anthers terminal, incumbent, cucullate; pollinia 8 in 2 groups, each group with a small front pair and a large hind pair, oblong-obovoid, connected by a granular mass at the apex; stigma triangular; ovary cylindric. Capsules oblong-ellipsoid, strongly ridged. Terrestrial perennial orchids with large fleshy purplish brown conic erect pseudobulbs. Stem erect, changing into pseudobulb after anthesis. Cataphylls green at flowering time. New shoot emerging from the base of a mature pseudobulb, consisting of two elements, the flowering branch and the leafy shoot. Leaves 3 or 4, apical to a pseudobulb, chartaceous, arching, plicate-venose. One species in Hong Kong, *A. sinense* Rolfe.

45. **Phaius** Lour. 鶴頂蘭屬 (FIGURE 51)

Flowers large, showy, in a terminal raceme; sepals and petals similar in size and color, free, spreading; lip undivided, trumpet-shaped, embracing the column, calcarate, the spur short and pointed, the apical margin wavy; column long, slightly enlarged and winged at the apical portion, without foot; clinandrium short, oblique, sinuate at the margin; rostellum subtruncate; anther attached to the margin of the clinandrium, incumbent, convex; pollinia waxy, 8, unequal, obovoid, slightly compressed laterally; connected at the apex by granular mass; stigma suborbicular; ovary obovoid-cylindric. Capsules ellipsoid, pendulous, rostrate, ridged. Terrestrial perennial orchids growing in shade of banks of streams, requiring dormancy in the dry cool season. Young shoot emerging from the base of an one-year old pseudobulb and changing into a pseudobulb after anthesis. Leaves 5 or 6, plicate-venose. Scape axillary to the second leaf from the bottom. One species with two varieties in Hong Kong; *P. tankervilliae* (Banks ex l'Herit.) Blume var. *tankervilliae* with sepals and petals brown within, widespread, and var. *veronicae* Hu & Barr. with sepals and petals lemon-yellow within, restricted in distribution.

FIGURE 50. *Acanthephippium sinense*: A. The habit sketch of a fruiting plant showing the initiation of a new shoot from a mature pseudobulb. Note the relative position of the leafy and the flowering shoots, and the ellipsoid capsule with the remains of old perianth. B. The habit sketch of a plant at the flowering stage with the leafy portion of the mother shoot removed, showing the relative positions of the flowering shoot and the new vegetative shoot, and the odd and urceolate lateral sepals. C. The lateral view of a flower with one-half of the odd sepal, a lateral sepal, and one sidelobe of the lip removed, showing the lip with small rectangular erect sidelobe and lamellate disc, the column with curved foot associated with a mentum, the helmet-like anther, and the stigma. D. The apical portion of the column showing the helmet-like anther and the stigma. E. Eight unequal pollinia in two groups of 2 pairs each.

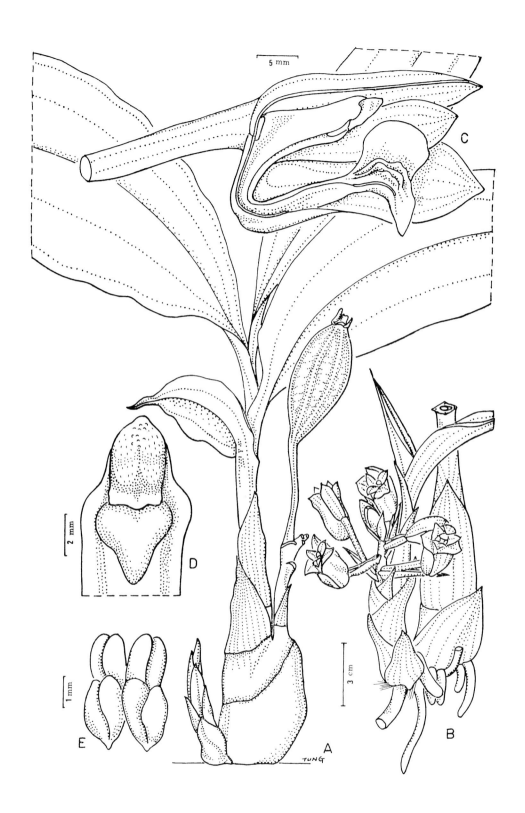

FIGURE 51. *Phaius tankervilliae*: A. The habit sketch of a flowering plant showing an old pseudobulb and the position of the scape. B. The upper portion of a scape showing 2 sterile bracts, the raceme with open flowers and flower buds, the position of the spur before and after resupination, and the trumpet-shaped lip embracing the column. C. A flower with the sepals, the petals, and one-half of the lip removed, showing the attachment of the lip to the base of the column, the papillose inner surface of the spur, and the disc of the lip, a glabrous callus, and the column slightly papillose on the sides and winged at the apex. D. The front view of an apical portion of the column showing the wings, the anther hairy on the sides, the subtruncate rostellum, and the large obovate stigma. E. A longitudinal section of the column with portion of the ovary attached, showing a deep cavity, the apical attachment of the anther cap, the position of the pollinia with the front pairs under the anther, and the hind pairs in the cavity, and the rostellum separating the stigma and the clinandrium. F. Eight unequal pollinia in 4 pairs with the front ones smaller than the hind ones. G. The habit sketch of a fruiting branch showing pendulous capsules with persistent column.

92

46. **Calanthe** R. Br. 蝦脊蘭屬 (FIGURE 52)

Flowers medium-sized, in a terminal raceme; sepals equal, free, spreading; petals almost similar as or slightly narrower than the sepals, spreading; lip adnate to the sides of the column, calcarate, 3-lobed, the spur slender, usually longer than the ovary, lateral lobes erect or spreading, midlobe truncate or bilobed, disc with cristate or lobed calli; column short, without foot; clinandrium membranaceous; rostellum triangular, acuminate, bifid; anther subterminal, incumbent, acuminate; pollinia 8, waxy, lanceolate, attenuate at the apex; viscidium linear or obovate; stigma transversely dumb-bell-shaped, visible from the sides of the rostellum, or U-shaped below the rostellum; ovary subcylindric. Capsules oblong-ellipsoid. Terrestrial perennial orchids, caespitose. Leaves large, plicate-venose, chartaceous. Scape axillary to the second or the fourth leaf of the current year's growth. Four species in Hong Kong.

Key to the Species

A. Flowers yellow, appearing with the young shoot in early spring; spur short; lateral lobes of the lip broad, suborbicular; midlobe truncate or shallowly bilobed.
  B. Flowering bracts deciduous; sepals yellow on both surfaces; spur clavate; midlobe of lip 2-lobed, with the base narrower than the apex.............*C. patsinensis* S. Y. Hu
  BB. Flowering bracts persistent; sepals tinged brown on the outside; spur curved, acute; midlobe of lip truncate, the base and the apex similar in width.........................
  ....*C. striata* (Sw.) R. Br.
AA. Flowers white or purple-red, appearing in the autumn on shoots with mature leaves; spur slender, equal to or longer than the ovary and the pedicel together; lateral lobes of the lip ovate or obovate-oblanceolate; midlobe of the lip obcordate or with narrow divergent lobes.
  B. Flowers white, in a crowded corymb-like raceme; spurs pointing downward; midlobe of lip with divergent narrow lobes............................*C. triplicata* (Willem.) Ames
  BB. Flowers purple-pink, in a loose raceme; spurs very long, pointing backward; midlobe of the lip obcordate.....................................*C. masuca* Lindl. var. *sinensis* Rendle

47. **Lycaste** Lindl. 麗仙蘭屬 (FIGURE 53)

Flowers large, showy, solitary, terminal to a scape; sepals subequal, spreading, the lateral ones oblique and shortly adnate to the column-foot at the base forming a short mentum; petals smaller than the sepals, close together with the upper margin slightly overlapping over the column; lip shorter than the sepals, 3-lobed, saccate, attached to the column-foot; lateral lobes erect, midlobe ovate, slightly narrowed at the base, disc with a broad and flat callus; column long, semiterete, arcuate, produced into a short foot; clinandrium short; rostellum truncate, emarginate; anther terminal, incumbent, strongly convex; pollinia 4 in 2 pairs, oblong or ovoid, attached to a slender stipe with a viscidium; ovary slender. Epiphytic perennial orchids. Rhizomes short, pseudobulbs close together, compressed ovoid, 4–8 cm. long, 3–5 cm. across the broad side. Leaves large, plicate-venose. Scape emerging from the base of the pseudobulb, with basal and median bracts, the flowering bract spathiform. One species introduced from Guatemala for ornamental purposes, *L. skinneri* Lindl.

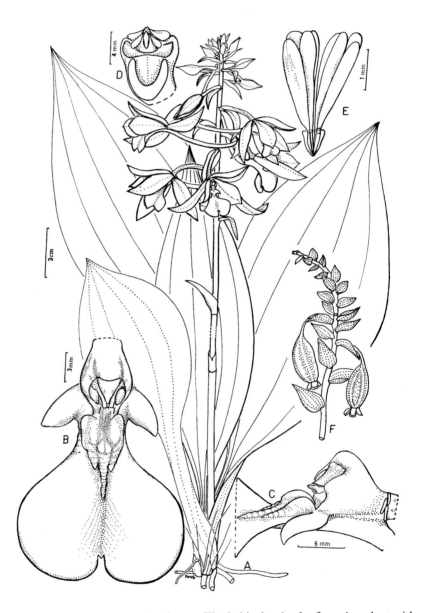

FIGURE 52. *Calanthe masuca* var. *sinensis*: A. The habit sketch of a flowering plant with portion of the scape shortened, showing the leaves and flowers with elongated spurs. B. The front view of flower with sepals and petals removed showing the column, the lip with small lateral lobes, an obcordate midlobe, and the callus with 3 cristate ridges, the anther, the rostellum, and the exposed portion of the stigma. C. The lateral view of the column and portion of lip showing the connection of these 2 organs, the anther, and the bifid rostellum. D. The front view of the column with the lip removed, showing the dumb-bell-like stigma with very thin portion beneath the bifid rostellum. E. Eight slender subequal pollinia in 4 pairs on a broad viscidium. F. The habit sketch of a fruiting branch showing the persistent bracts, and the pendulous capsules with the remains of the perianth.

FIGURE 53. *Lycaste skinneri*: The habit sketch of a flowering plant showing the root, the compressed-ovoid pseudobulb, the leaves below and above the pseudobulb, the position of the scape, the solitary flower with the petals much smaller than the sepals, the trilobed lip with a callus on the disc, the anther, and the column.

48. **Geodorum** Jacks. 地寶蘭屬 (FIGURE 54)

Flowers rather small, crowded in a nutant raceme; sepals and petals subequal, free, ovate, spreading; lip undivided, attached to the foot of the column, concave, the base ventricose-subsaccate, not spurred, with a thick papillose curved callus above the sac, disc keeled; column short, broad, subterete, with a short foot fused with the lip; clinandrium broadly ovate, apiculate; rostellum truncate, fleshy; anther subglobose, incumbent, with 2 basal flaps; pollinia 2, waxy, globose and profoundly saccate on the back, attached to a slender hyaline stipe with a small viscidium; stigma obovate-suborbicular; ovary oblong. Capsules oblong-ellipsoid, nutant, strongly rostrate. Terrestrial perennial orchids. Rhizomes tuberous, ovoid or compressed spherical, with concentric rings. New shoots emerging from the base of the pseudobulb. Leaves 3 or 4, oblong-linear, plicate-venose, deciduous. Scape emerging from an one-year old rhizome simultaneously with the leafy shoot, the apex bending at anthesis, becoming erect afterward. One species in Hong Kong, *G. densiflorum* (Lam.) Schlechter, requiring dormancy in the cool dry season.

49. **Cymbidium** Swartz 蘭屬 (FIGURE 55)

Flowers moderately large, in a loose raceme; sepals equal, free, spreading; petals similar to or slightly smaller than the sepals, spreading or proximate and over the column; lip attached to the base of the column, 3-lobed, the base fleshy, concave, lateral lobes erect, partially covering the column, midlobe strongly recurved, disc 2-keeled and deeply canaliculate, rarely tubercular; column long, curved, without foot; clinandrium oblique-truncate; anther terminal, incumbent, convex or subglobose; pollinia 2, folded, waxy, in several species separated into four readily, on a broad stipe widened to a viscid base; stigma suborbicular; ovary cylindric, striate-sulcate. Capsules fusiform or oblong-ellipsoid, strongly rostrate. Perennial caespitose herbs, terrestrial or epiphytic. Roots wirelike. Rhizomes short and thick, covered by cataphylls and enlarged into conic pseudobulbs. Leaves 3–5, persistent, petiolate, linear or lanceolate. Scape emerging from the axil of a cataphyll or rarely of a leaf, with imbricate basal scales and cauline sheaths; flowers often very fragrant. A favorite genus for Chinese orchidophiles, many cultivars in private gardens of Hong Kong; 4 spontaneous species observed in woods and bamboo groves, one in cultivation.

Key to the Species

A. Leaves lanceolate; cataphylls many, imbricate.
    B. Petals spreading; short-day flowers, appearing in December and January; capsules parallel to the rachis...*C. maclehoseae* S. Y. Hu
    BB. Petals proximate, forming a roof over the column; long-day flowers, appearing in May–June; capsules reclining, forming a 45° angle to the rachis....*C. lancifolium* HK.
AA. Leaves linear; cataphylls few, almost parallel.
    B. Plants erect; flowers 5–7 cm. across; disc keeled and canaliculate.
        C. Scapes 1 m. or more high; leaves 3 cm. or more wide; short-day flowers, appearing in January–February, faintly fragrant.
            D. Flowers greenish, appearing purplish black due to many longitudinal purple lines, fragrant, opening in winter........................*C. sinense* (Andr.) Willd.
            DD. Flowers yellowish green, without any purplish lines or spots, opening in winter, slightly fragrant.................*C. sinense* var. *albo-jucundissimum* (Hay.) Fukuy.
        CC. Scapes 20–40 cm. long; leaves 8–15 mm. wide; long-day flowers, appearing in June–October, strongly fragrant.
            D. Petals each with a median-longitudinal purplish-red line; lip maculate with irregular bars, patches and spots; flowering in autumn....*C. ensifolium* Swartz
            DD. Petals without median longitudinal red line; lip uniformly pale yellow; flowering in summer (cultivated).............................*C. xiphiifolium* Lindl.
    BB. Plants epiphytic, pendulous, growing on walls or trees; flowers 3–4 cm. across; disc with tubercular calli...*C. pendulum* (Roxb.) Swartz

---

FIGURE 54. *Geodorum densiflorum*: A. The habit sketch of a group of the pseudobulbs of six years' growth showing the relative positions of the scape and vegetative shoot. B. The front view of a flower showing the relative positions of the sepals, petals, lip, and column. C. The lateral view of a flower showing the sepals, petals, and the undivided lip with very short apical portion. D. The same with one-half of dorsal sepal, 1 lateral sepal, 1 petal, and one-half of the lip removed, showing the undivided trowel-shaped lip with front keels and back callus, and the column with the anther and stigma. E. An apical portion of the column with the anther cap pushed back slightly, showing the 2 globose pollinia on a short stipe, the rostellum, and the stigma. F. The back view of two cleft pollinia on a stipe. G. Abaxial view of the lip. H. The habit sketch of a fruiting branch showing the pendulous oblong capsules with persistent beak-like column. I. The lateral view of an undivided lip with keeled disc.

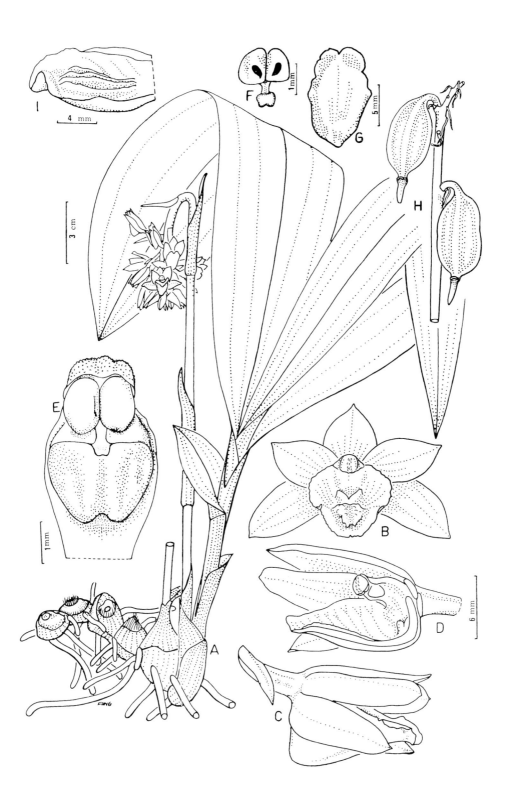

50. **Oncidium** Swartz 瘤唇蘭屬 (Figure 56)

Flowers showy, in terminal panicles; sepals subequal, spreading, free or the lateral ones shortly connate at the base; petals similar or slightly larger than the odd sepal; lip attached to the base of the column, 3-lobed, disc cristate, lateral lobes small, subrotundate, midlobe spreading, suborbicular, emarginate; column short, stout, extended into petaloid auricles on both sides of the stigma; clinandrium oblique erect; anther terminal, incumbent, convex, cucullate, imperfectly 2-locular; pollinia 2, waxy, obovoid, on short stipe with viscidium. Capsules fusiform. Epiphytic orchids with very short stem furnished with equitant cataphylls and pseudobulbs. Leaves 2 or 1, terminal to a pseudobulb, coriaceous. Inflorescences paniculate; peduncles apical to pseudobulbs, bearing several sterile bracts. Flowers pedicellate, the yellow sepals and petals barred purplish-red. Native of tropical America, 1 species cultivated in Hong Kong, *O. varicosum* Lindl.

51. **Odontoglossum** H.B.K. 齒舌蘭屬 (Figure 57)

Flowers showy, in an axillary raceme; sepals subequal, free, spreading, oblong, lanceolate, or rarely ovate; petals similar to or broader than the sepals; lip attached to the base of the column, undivided, narrowed at the base and bent, the narrow base parallel to the column and the widened apical portion deflexed, suborbicular or reniform, emarginate, the disc with two calli; column long, the upper portion petaloid and dentate, the lower portion subterete; clinandrium truncate; anther terminal, incumbent, semiglobose; pollinia 2, ovoid, sulcate on the side; stipe linear; viscidium ovate. Capsules oblong, rostrate. Epiphytic orchids with short stems bearing equitant leaves and one to several pseudobulbs each terminated by 1 or 2 coriaceous, oblong-lanceolate leaves. Native of tropical America, much used by hybridizers for bigeneric combinations, with X *Odontioda* (*Odontoglossum* X *Cocblioda*) and X *Odontocidium* (*Odontoglossum* X *Ondidium*) being the well-known productions; 2 species cultivated in Hong Kong.

Key to the Species

A. Flowers 10 cm. or more across, yellow with reddish brown bands and spots; lip suborbicular; pseudobulbs 8–10 cm. long (Guatemala)............................*O. grande* Lindl.
AA. Flowers 8 cm. or less across, white and tinged rose, the lip violet or deep rose; lip reniform and emerginate; pseudobulb 10–15 cm. long (Mexico-Guatemala) ....
....*O. pendulum* Batem. (*O. citrosmum* Lindl.)

Figure 55. *Cymbidium ensifolium*: A. The habit sketch of a flowering plant showing the caespitose habit, the evergreen leaves, the position of the scape, and the various views of the flowers. B. The front view of a flower showing the bract, ovary, sepals, petals, lip, column, and the anther. C. The same with all sepals and petals removed, showing the lip with a median longitudinal groove, the erect sidelobes, the column, and the anther. D. The same with one-half of the lip removed, showing the thick disc and the curvature of the midlobe of the lip, the column, and the anther. E. The front view of the column showing the anther, the rostellum, and the stigma. F. Two pairs of compressed pollinia on a stipe attached to a broad viscidium. G. The habit sketch of a fruiting branch showing two erect rostrate capsules parallel to the axis of the rachis.

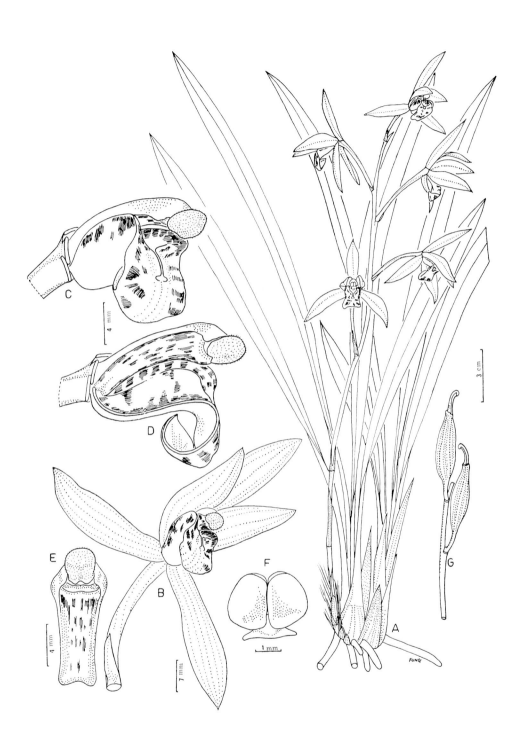

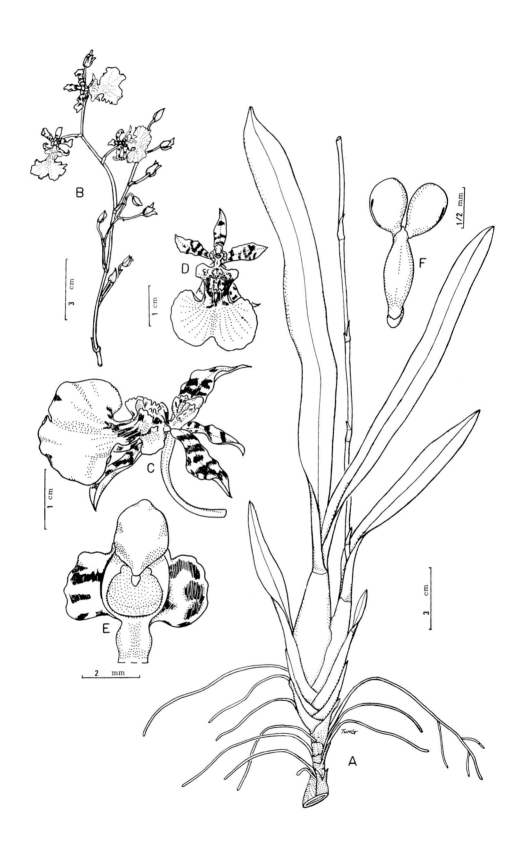

FIGURE 57. *Odontoglossum grande*: The habit sketch of a flowering plant showing the pseudobulbs and the axillary raceme. (Redrawn from plate 24 of James Bateman's The Orchidaceae of Mexico and Guatemala, London, 1840.)

FIGURE 56. *Oncidium varicosum*: A. The habit sketch of a plant showing the roots, the stem with equitant cataphylls and leaves, the pseudobulbs terminated by leaves or a leaf, and a scape. B. A portion of a panicle with flowers and buds. C. The lateral view of a flower showing the relative positions of the barred sepals and petals, the lip with cristate disc, the column, and the anther. D. The front view of the same. E. The front view of the column showing the petaloid wings, the anther, and the large stigma. F. Two obovoid slightly cleft pollinia on an elliptic stipe with small viscidium.

52. **Miltonia** Lindl. 米頓蘭屬 (FIGURE 58)

Flowers large, showy, 3–4 on a loose axillary raceme; sepals free, equal, spreading, ovate-oblong; petals slightly broader than the sepals, spreading, slightly upright; lip undivided, attached to the base of the column, large, flat, emarginate or shallowly bilobed at the apex, narrowed and auriculate at the base; column short, without foot, the apical portion winged and auriculate; clinandrium short, truncate or 2-lobed; rostellum truncate; anther terminal, incumbent, convex; pollinia 2, waxy, ovoid, entire or sulcate, attached to an oblong or obovate stipe with a small viscidium; stigma large, covering the front of the column; ovary cylindric. Epiphytic perennial orchids, stem stout, suberect, crowded with the equitant petioles of the leaves and terminated by a pseudobulb with 1 leaf. Leaves coriaceous, elongate-oblong, slightly narrowed at the base. Scape axillary to a leaf below the pseudobulb. Flowers on elongate pedicels; bracts spathiform. One cultivar introduced from tropical America, *Miltonia* "Belt Field," for ornamental purposes.

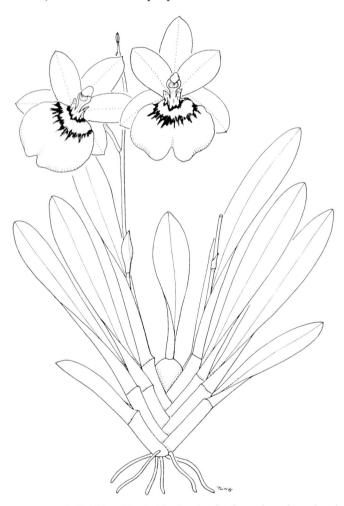

FIGURE 58. *Miltonia "Belt Field"*: The habit sketch of a flowering plant showing the roots, equitant leaves, the axillary scapes, the large colorful flowers with oblong sepals and petals, the large lip auriculate at the base, and the column with anther and stigma.

53. **Thrixspermum** Lour. 白點蘭屬 (FIGURE 59)

Flowers showy, ephemeral, in a short axillary raceme; sepals equal, free, spreading; petals slightly smaller; lip 3-lobed, saccate, adnate to the base of the column, spur conic, lateral lobes erect, obtuse, midlobe triangular-ovate, obtuse, disc with a callus; column short, semiterete, with fleshy arms; clinandrium oblique; rostellum fleshy, horse-shoe-shaped; anther terminal, incumbent, convex; pollinia 4 in 2 pairs, unequal, the front one of each pair smaller, obovoid-oblong, on a white viscidium with a recurved ovate membrane resting over the horse-shoe-shaped rostellum; stigma suborbicular; ovary obovoid-cylindric. Capsules oblong. Epiphytic orchids. Aerial roots long and branched. Leaves oblong-elliptic, distichous. Scapes short, subsessile, many on a shoot, with imbricate persistent scales. One species in Hong Kong, *T. centipeda* Lour. (*T. hainanensis* [Rolfe] Schlechter).

54. **Arachnis** Blume 蜘蛛蘭屬 (FIGURE 60)

Flowers moderately large, showy, in axillary racemes; sepals and petals almost equal in size and shape and similar in color, free, spreading, narrow, widening to the tips, recurved at the edge, the petals curved downward toward the apex; lip movable, attached to the column-foot by a thin elastic hinge, neither saccate nor calcarate, 3-lobed, the lateral lobes almost rectangular, erect, parallel to the sides of the column, their ends diverging, midlobe fleshy, with a median keel, interrupted at the base, with a fleshy callus below the apex, disc erect, forming a right angle with the midlobe, gibbous at the apex, and with the tip of the sac pointing backward; column short, stout, with a short foot; clinandrium truncate; rostellum truncate; anthers terminal, incumbent, the apex obtuse; pollinia 4 in 2 pairs, each pair obovate-globose, stipe deltoid-lanceolate, the broad base curved and the apex recurved; stigma concaved. Epiphytic orchids. Roots stout, green. Stems terete. Leaves distichous, ovate-oblong, coriaceous-carnose, bilobed at the apex. Native of tropical southeastern Asia, 2 species cultivated in Hong Kong for ornamental purposes.

Key to the Species

A. Internodes 4–10 cm. long; leaves 15–17 cm. long, 5 cm. wide; sepals and petals yellow green, maculate with dark purple-red bars; lip red, the midlobe broad with low keel, the inside of the lateral lobes with numerous fine red bars............*A. flos-aeris* (L.) Reichb. f.

AA. Internodes 2.5–5 cm. long; leaves 10–12 cm. long, 2.5 cm. wide; sepals and petals lemon yellow, white at the base, maculate with purple-red bars; lip yellow, the midlobe narrow, with high keel, the lateral lobes uniformly light yellow, without bars.........................
....*A. maingayi* (HK. f.) Schlechter

55. **Renanthera** Lour. 火焰蘭屬 (FIGURE 61)

Flowers moderately large, showy (red), in axillary panicles; sepals free, spreading, unequal, the odd one erect, linear-oblanceolate, the lateral ones proximate to each other, longer and broader than the odd sepal, widened obliquely at the apical half; petals linear, slightly shorter than the odd sepal; lip short, attached to the base of the column, 3-lobed, saccate, the sac conic, lateral lobes erect, rectangular, midlobe small, recurved, disc white, with 2 white high oblong calli; column short, semiterete, without a foot; clinandrium small; rostellum truncate; anther terminal, ovate, convex; pollinia 4, subequal, waxy, ovoid; stipe broad, with viscidium; stigma concaved; ovary cylindric. Capsules cylindric. Tall epiphytic orchid up to 2 m. high. Stems monopoidal. Leaves distichous, fleshy, 6 cm. long, 3 cm. wide, obliquely 2 lobed at the apex. One species native to Kwangtung and the adjacent countries to the south, cultivated in Hong Kong for ornamental purposes, *R. coccinea* Lour.

56. **Diploprora** Hook. f. 蛇舌蘭屬 (FIGURE 62)

Flowers small, in a short axillary spike, opening one at a time; sepals subequal, free, spreading; petals slightly narrower than the sepals, spreading; lip 3-lobed, attached to the base of the column, concave, without a foot or spur, lateral lobes erect, obtuse, midlobe narrow, the apex extending into 2 filiform lobes, disc flat, fleshy; column short, oblong, clinandrium truncate; rostellum ovate, obtuse; anther semiglobose; pollinia 4 in 2 pairs, unequal, the front ones larger, obovoid, attached to a stipe with an ovate viscidium; stigma large, covering the entire front of the column; ovary cylindric. Capsules oblong, the apex crowned with persistent sepals. Epiphytic monopodial orchids growing on boulders in shade of trees along streams or on cliffs of peaks with frequent cloud. Roots string-like, branched. Leaves leathery, distichous. Scapes several, axillary, short, with basal scales; flowering bracts persistent. One species in Hong Kong, *D. championii* (Lindl.) Hook. f.

57. **Phalaenopsis** Blume 蝶蘭屬 (FIGURE 63)

Flowers large, showy, in axillary racemes or panicles; sepals subequal, spreading and equally spaced, oblong; petals suborbicular and contracted at the base; lip constituting a continuation of the column-foot, without a hinge, ecalcarate, with a large and fleshy cleft callus at the base, 3-lobed, the lateral lobes erect, narrowed at the base, the midlobe lanceolate or rhomboid-lanceolate, the apex often furnished with 2 antennae and a notch below the blade; column stout, semiterete, the foot well developed; rostellum lanceolate, acuminate, 2-lobed at the apex; anther ovate-acuminate; pollinia 2, obovoid-subspherical, hollow on the back, attached to a spathulate stipe; viscidium scale-like, obcordate; stigma concaved, suborbicular. Capsules clavate. Epiphytic monopodial orchids. Roots cord-like, numerous. Stems short. Leaves close together, distichous, oblong-linear. Scape axillary. Native of tropical southeastern Asia, several species cultivated in Hong Kong.

Keys to Species and Cultivars Observed

A. Lip without filiform antenna; flowers medium-sized, 2–3 cm. across.
    B. Flowers normal, with bilateral symmetry; the inner perianth-segments consisting of 2 pink-purple petals and a 3-lobed lip; column long and slender, with a linear rostellum longer than the ovate-acuminate anther (Philippines)..*P. equestris* Reichb. f. (= *P. rosea* Lindl.)
    BB. Flowers highly modified, with radial symmetry; all the inner perianth segments petaloid, unguiculate, 3-lobed, ovate, acuminate, with purplish pink stripes and a callus on the disc; without column and the associated organs (Philippines, hybrid origin)...*Phalaenopsis* "Star of Leyte"
AA. Lip with 2 antennae; flowers large, showy, 5–7 cm. across.
    B. Flowers white; leaves 10–18 cm. long, 3.5–5 cm. wide; antennae recurved (Melasia)..*P. amabilis* (L.) Blume
    BB. Flowers purplish-pink; leaves 20–35 cm. long, 6–10 cm. wide; antennae divergent (Philippines)...*P. schilleriana* Reichb. f.

FIGURE 59. *Thrixspermum centipeda*: A. The habit sketch of a flowering plant showing the monopodial stem, numerous roots, and the lateral inflorescences. B. The lateral view of a flower with 1 sepal and 1 petal removed, showing the saccate lip with erect lobes, and the column. C. The front view of a flower with all the sepals and petals removed, showing the lip with the lateral lobes spread out and a callus on the disc, the column with thick fleshy arms, the anther, the viscidium, and the stigma. D. The apical portion of the column with the anther removed, showing the thickened wings, the viscidium covering the fleshy rostellum and the ovate hyaline membrane. E. The back view of four unequal pollinia in 2 pairs. F. The front view of four unequal pollinia on a narrow viscidium with a recurved ovate membrane. G. The back view of the viscidium and the ovate hyaline membrane.

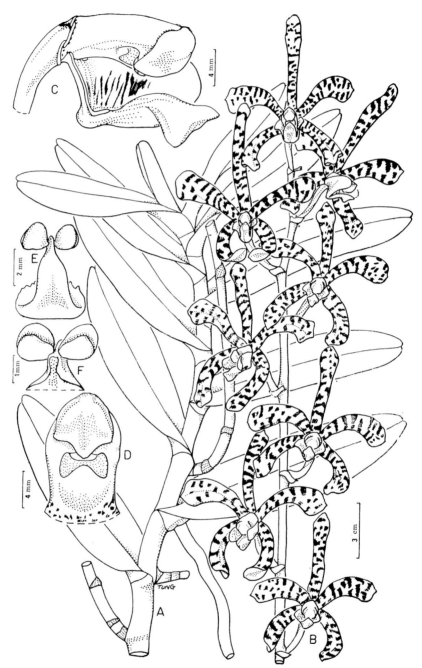

FIGURE 60. *Arachnis flos-aeris*: A. The habit sketch of a leafy shoot showing the monopodial stem with stout aerial root and axillary scape. B. A portion of the raceme showing the flowers with narrow maculate more or less curved sepals and petals, the small 3-lobed lip, and the column. C. The lateral view of a flower with all the sepals and petals and one lateral lobe of the lip removed, showing the movable lip with a rectangular sidelobe, a fleshy midlobe, a callus below the apex, and a gibbous bent disc. D. The front view of a column showing the anther, stigma, and the viscidium. E. The front view of the 4 pollinia on a stipe with a bent viscidium. F. The back view of the paired pollinia on the stipe.

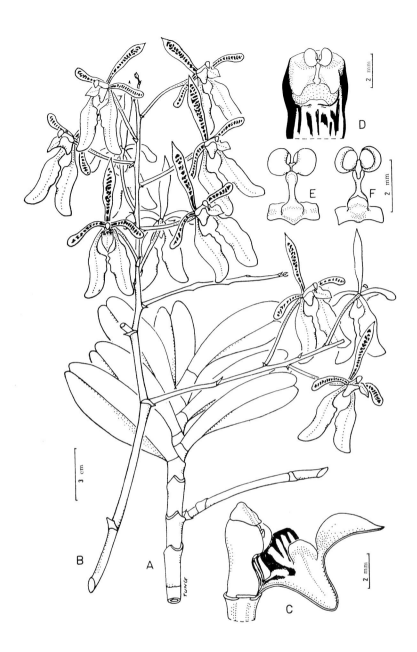

FIGURE 61. *Renanthera coccinea*: A. The habit sketch of a leafy shoot with monopodial growth, showing the distichous leaves and the position of the inflorescence. B. The habit sketch of a panicle showing the flowers with linear odd sepal, broader lateral sepals, shorter petals, short lip, and slender column. C. The lateral view of a flower with the sepals, petals, and one-half of the lip removed, showing the lip with a conic spur, an erect lateral lobe striped on the inner surface, a high callus of the disc, and one-half of the ovate midlobe, the subterete column with fleshy arms at the apex, and the ovoid anther. D. The front view of the column with the anther cap removed, showing the truncate small clinandrium, the pollinia on a broad stipe, and the concaved stigma with a shiny gland at the middle of the lower margin. E. The front view of the pollinia on a broad stipe attached to the rectangular membranous viscidium. F. The back view of the same showing the unequal sizes of the pollinia.

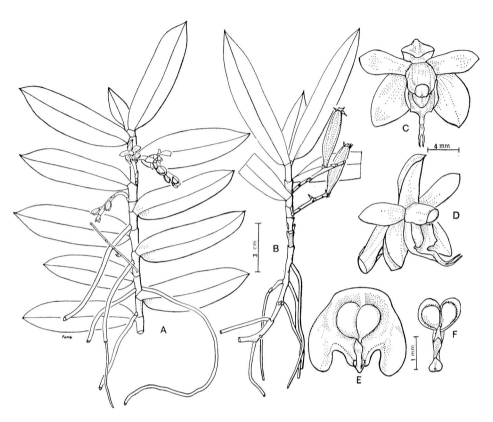

FIGURE 62. *Diploprora championii*: A. The habit sketch of a flowering plant showing the monopodial stem, and the positions of the aerial roots and the racemes. B. The habit sketch of a fruiting plant showing the cylindric capsules with persistent perianth segments. C. The front view of a flower showing the sepals, petals, lip with narrow midlobe forked at the apex, and the broad column with the anther at the apex. D. The lateral view of a flower showing the sepals, the petals, the column, and the lip with erect lateral lobes, the forked narrow mid-lobe, and a flat thick disc. E. The apex of the column with the anther cap removed, showing the pollinia, stipe, viscidium, and the rostellum. F. The back view of four unequal pollinia in 2 pairs showing the attachment to the stipe and the viscidium.

58. **Microcoelia** Lindl. 細根蘭屬 (FIGURE 64)

Flowers small, in loose axillary racemes, non-resupinate; sepals and petals subequal, similar in sizes and color, free, spreading; lip undivided, attached to the base of the column, calcarate, the spur conic, lateral lobes obsolete, midlobe ovate, acute; column short, stout, without wings and foot; clinandrium truncate; rostellum

FIGURE 63. *Phalaenopsis amabilis*: A. The habit sketch of a flowering plant showing numerous aerial roots clinging to a support, the monopodial stem with large leaves distichously arranged, the position of scapes, the front, lateral and top views of flowers, and flower buds. B. A flower with all the sepals and petals and a lateral lobe of the lip removed, showing a small bract, the pedicel and ovary, the 3-lobed lip with a cleft callus on the disc and two filiform antennae at the apex, the stout column with well-developed foot, the anther, and the beaked rostellum. C. The front view of the column showing the ovate and acuminate anther, the bifid rostellum, the apical portion of the stipe with a viscidium below the rostellum, and the stigma. D. The front view of the spathulate stipe with two obovoid-subspherical pollinia and an obcordate viscidium. E. The back view of the pollinia showing the hollow back.

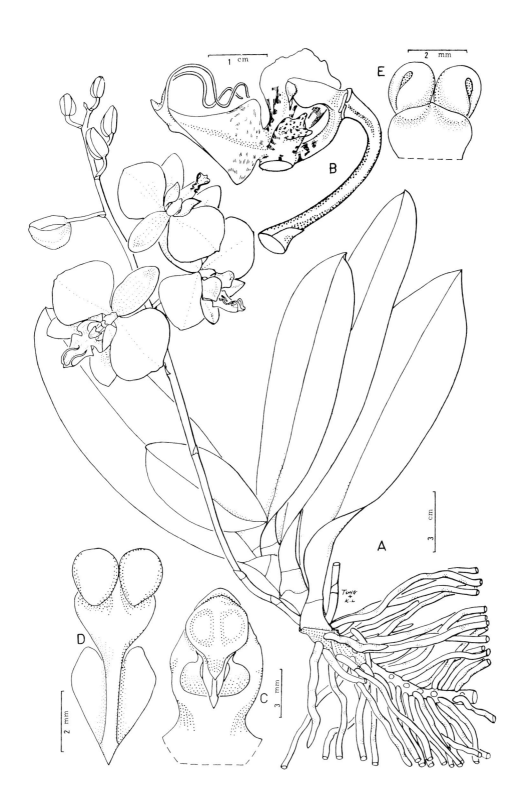

ovate-acuminate, bifid; anther terminal, incumbent, convex; pollinia 2, waxy, obovoid, apical to a filiform stipe; viscidium below the lobes of the rostellum; stigma suborbicular; ovary oblong. Epiphytic aphyllous orchids. Roots numerous and long, greenish gray. Stem very short, densely covered by acuminate bracts. Scapes filiform, bracts minute, persistent. Flowers 3 mm. long, creamy-white, the lip yellowish green, with a brownish spur. One African species introduced into cultivation in Hong Kong, *M. guyoniana* (Reichb. f.) Summerhayes.

59. **Acampe** Lindl. 脆花屬 (FIGURE 65)

Flowers rather small, in axillary small panicles; sepals and petals similar in color and size, free, spreading, with transverse bars; lip 3-lobed, attached to the base of the column, slightly saccate at the base, papillose-velvety, fleshy; column short, stout, without wing or foot, with a tooth on each side of the apex; clinandrium truncate; rostellum obtuse; anther semiglobose; pollinia 4 in 2 globose pairs, waxy, on a slender stipe; viscidium ovate; stigma large, suborbicular; ovary cylindric. Capsules cylindric, erect, erostrate. Large fleshy epiphytic orchids with monopodial growth. Roots stout, long, branched, clinging on damp rocks under trees along streams. Leaves band-like, 20 cm. long, 5 cm. wide. Scape shortly pedunculate. Flowers crowded near the apex of the flowering branch. One species in Hong Kong, *A. multiflora* (Lindl.) Lindl.

60. **Vanda** R. Br. 萬帶蘭屬 (FIGURE 66)

Flowers large, showy, in axillary racemes; sepals and petals similar in size and color, spreading, narrowed at the base; lip attached to the short column-foot, calcarate, trilobed, lateral lobes erect, rotundate, the midlobe rotundate or oblong, the apex cleft, disc with longitudinal ridges and sulcate, with 2 small calli at the base before the entrance of the conic spur; column stout, short; clinandrium truncate; anther terminal, incumbent, subglobose, the apex acute, acuminate or obtuse; pollinia 2, deeply cleft on the back; stipe spathulate, viscidium large, recurved below the subtruncate rostellum; stigma suborbicular, deeply concave; ovary elongated, 6 furrowed, and sharply 6-ridged, slightly twisted. Capsules cylindric, prominently ridged. Epiphytic orchids with monopodial growth, stems stout or cylindric and slender. Leaves flat or terete, distichous. Inflorescences racemose or paniculate; bracts short. Native of southeastern Asia, many hybrids cultivated in private gardens of Hong Kong. Garay in 1974 transferred several species of *Vanda* to *Papilionanthe* Schlechter. Here the traditional treatment of the group as appeared in Holttum's Orchids of Malaya is adopted.

Key to the Species Observed

A. Flowers of several colors; leaves flat; scape with 5–10 flowers.
  B. Leaves 12–15 cm. long, 3 cm. wide; flowers 8–10 cm. across, light bluish or pinkish purple; petals and sepals subrotundate, 3 cm. across; midlobe of lip narrowed at the base, subrotundate at the apex; disc with 5 keels (Eastern Himalayan Region). ....*V. coerulea* Griff. ex Lindl.
  BB. Leaves 30–40 cm. long, 2–3 cm. wide; flowers 5–6 cm. across, white with yellowish tint and with numerous purple spots, the lip violet purple; petals and sepals oblanceolate-oblong, 1–1.5 cm. across; midlobe of lip panduriform, widened at the base, oblong at the apical half; disc with 3 keels (Native of Java)..........*V. tricolor* Hooker
AA. Flowers pure white; leaves terete; scape with 3–5 flowers (Native of Burma)............... ....*V. teres* Lindl.

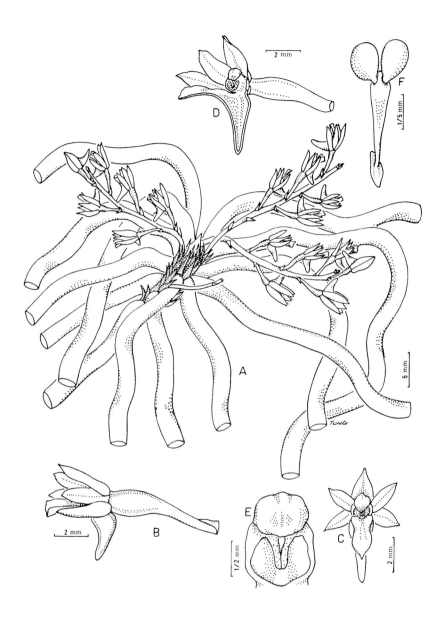

FIGURE 64. *Microcoelia guyoniana*:   A. The habit sketch of a flowering aphyllous plant showing the numerous terete roots, the short stem with leaves reduced to imbricate scales, the axillary position of the racemes, the various views of the non-resupinate flowers.   B. The lateral view of a flower showing a small bract, the sepals, petals, and the calcarate lip.   C. The front view of the same showing the sepals, petals, lip, column, anther, rostellum, and the stigma.   D. The lateral view of a flower with a sepal, a petal, and one-half of the lip removed, showing the spur, the short column, anther, bifid beaked rostellum, and the stigma.   E. The front view of an apical portion of the column showing the subglobose anther, a portion of the stipe, the bifid rostellum, and the stigma.   F. Two obovoid-subglobose pollinia on a stipe with elliptic viscidium.

FIGURE 65. *Acampe multiflora*: A. The habit sketch of a flowering shoot showing the sympodial stem with stout aerial roots, distichously arranged band-like leaves, and the lateral position of the paniculate inflorescence. B. The front view of a flower showing the symmetrical arrangement of the sepals and petals, and the undivided lip with velvety surface. C. The lateral view of a flower with a petal and a lateral sepal removed, showing the small bract, the short pedicel and ovary, the short stout column, the subglobose anther, and the slightly saccate lip. D. The same, with the anther cap and a portion of lip removed, showing the thick, fleshy disc of the lip, papillose on the inside, the pollinia and stipe, and a very large stigma. E. The front view of 4 pollinia in two globose pairs attached to a stipe and a viscidium. F. The habit sketch of a fruiting branch showing the erect cylindric capsules with portions of the persistent perianth.

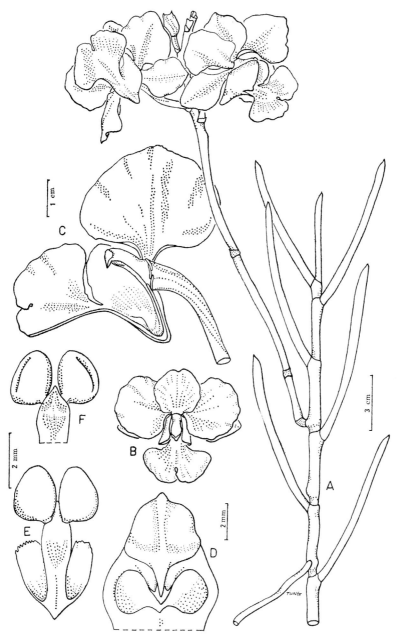

FIGURE 66. *Vanda teres*: A. The habit sketch of a flowering shoot showing the monopodial stem, the aerial root, the terete leaves, and the axillary scape with the lateral view and the lower front view of two flowers and a bud. B. The top front view of a flower showing sub-rotundate odd sepal and petals, the bilobed midlobe of the lip, and the column. C. The lateral view of a flower with the lateral sepals, the petals, and one-half of the lip removed, showing the erect lateral lobe, a callus, and the saccate base of the lip, the column, anther, and the stigma. D. The apical portion of the column showing the anther with rostrate apex, the bifid rostellum, and the stigma. E. The front view of two subglobose pollinia on a stipe wtih a large obcordate viscidium. F. The back view of a portion of the stipe showing the cleft pollinia.

61. **Gastrochilus** D. Don 肚唇蘭屬 (Figure 67)

Flowers rather small, non-resupinate, few on an axillary raceme; sepals and petals subequal, free, spreading in form of a fan; lip undivided, attached to the base of the column, broadly saccate, the apex transversely dilated and bilobed, the sac smooth inside, without callus or keel; column short, stout, winged; clinandrium truncate; rostellum subulate, bifid; anther terminal, incumbent, convex, acuminate at the apex; pollinia 2, subglobose in appearance from the front, flattened and with a central depression on the back; stipe slender; viscidium bifid; stigma suborbicular; ovary cylindric. Capsules oblong. Epiphytic orchids with monopodial growth. Roots numerous, thick and long. Stems rather short, covered by roots and remains of degenerated leaves. Leaves rather small, linear-lanceolate, wavy, acute. Scapes short, with 2–3 flowers. One species, from the cliff of Victoria Peak, *G. holttumianus* Hu & Barretto.

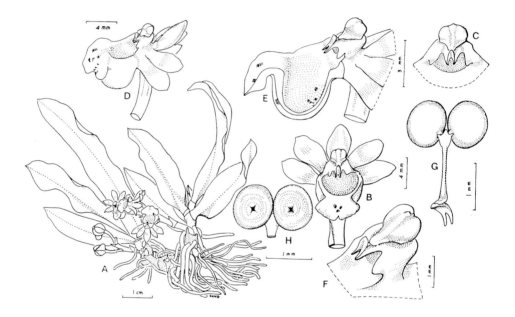

Figure 67. *Gastrochilus holttumianus*: A. The habit sketch of two flowering plants showing two shoots with roots, distichous leaves, lateral scapes, and non-resupinate flowers. B. The front view of a flower showing the well-spaced sepals and petals, the lip, column, anther, rostellum, and the bifid viscidium. c. The front view of a column showing the anther, the rostellum, and the forked viscidium. D. The lateral view of flower showing the ventricose lip, the column, the anther, and the position of the viscidium. E. The same with portion of the sepals and petals and one-half of the lip removed, showing the smooth inner surface of the sac of the lip. F. The lateral view of a portion of the column showing the anther, the attachment of the stipe to the viscidium, the rostellum, and the stigma. G. The front view of the pollinia showing the stipe, the attachment of the viscidium, and the 2 pollinia appearing spherical from the front. H. The back view of the pollinia showing a flat appearance with a depression in the middle.

62. **Robiquetia** Gaud. 露壁蘭屬 (FIGURE 68)

Flowers rather small, in an axillary panicle; sepals free, spreading, the odd one concave; petals slightly smaller than the sepals; lip 3-lobed, calcarate, attached to the base of the column; spur enlarged and rotundate at the apex, with a horizontal flap on the back wall near the entrance and a keel on the front wall, lateral lobes erect and truncate, midlobe fleshy, ovate; column short and broad, produced into two arms at the apex; clinandrium elongated; rostellum produced below the tip of the anther; anther terminal, ovate, the apex beaked; pollinia 4 in 2 pairs, on a short stipe attached to a slender caudicle enlarged, curved, and grooved at the pollinian end and with a small viscidium at the rostellum end; stigma suborbicular; ovary cylindric. Capsules fusiform. Epiphytic orchids with monopodial growth. Roots long, strong and branched. Stems slender. Leaves distichous, oblong, erose at the apex. Peduncles as long as the leaves, with basal and median sterile bracts. One species in Hong Kong, growing on cliffs, in shade, along streams, *R. succisa* (Lindl.) Seidenf. & Garay.

63. **Cleisostoma** Blume 封口蘭屬 (FIGURE 69)

Flowers rather small, in an axillary raceme or panicle; sepals subequal, free, spreading, the lateral ones attached to the column-foot; petals slightly smaller, spreading; lip 3-lobed, calcarate, attached to the base of the column, spur large, irregular-cylindric, obtuse, with a callus on the back wall at the entrance and a longitudinal keel on the front wall running down from the base of the midlobe, lateral lobes erect, over the sides of the sac, triangular, acute, midlobe fleshy, straight and triangular, or curved upward and with a callus at its base; column stout, semiterete, with a stout foot; clinandrium truncate, the apex 2-toothed; rostellum concave and U-shaped or subulate; anther terminal, incumbent, the apex truncate or acuminate; pollinia 4 in 2 pairs, waxy; stipe short compressed and enlarged to a horse-shoe-shaped base and viscidium, or slender and with a small viscidium; stigma orbicular; ovary obovate-cylindric. Capsules cylindric, with persistent column. Epiphytic monopodial orchids growing on exposed boulders. Roots wirelike, branched. Stems slender, terete. Leaves terete, lanceolate, or band-like. Three species in Hong Kong.

Key to the Species

A. Leaves terete; column foot papillose and barbate; midlobe of lip triangular-sagittate
....*C. teres Garay*

AA. Leaves flat, linear-lanceolate or band-like; column foot not barbate; midlobe of lip curved upward.
    B. Inflorescences paniculate, the peduncles similar to the leaves in length; leaves band-like, the apex retuse or bilobed; anther truncate-serrate at the apex; back callus of spur not meeting the front keel; stipe of pollinia short and broad, horse-shoe-shaped at the base...*C. paniculata* (Hance) Garay
    BB. Inflorescences racemose, the peduncles many times shorter than the leaves; leaves lanceolate, the apex acuminate; anther acuminate at the apex; back callus of spur connivent with the front keel; stipe of pollinia slender, obtuse at the apex............
....*C. fordii* Hance

64. **Ornithochilus** Wall. ex Lindl. 羽唇蘭屬 (Figure 70)

Flowers rather small, in an axillary panicle; sepals subequal, spreading, the odd one oblong, erect, the lateral sepals ovate and narrowed at the base, slightly twisted and attached to the column-foot forming a short mentum, reflexed and pointing upward; petals oblong, smaller than the sepals; lip much larger than the sepals, attached to the column-foot, shortly unguiculate, 3-lobed, calcarate, the spur slightly curved forward, the lateral lobes narrow, erect and slightly incurved, midlobe obcordate, strongly inflexed, bilobed, the lobes fimbriate, the base furnished with a velvety-papillose flap hanging over the orifice of the spur; column short, terete, with a short foot, oblique and papillose along the margin of the stigmatic cavity; clinandrium small, erect on the back; rostellum bifurcate, the prongs long and curved downward at the apical portion; anther terminal, helmet-shaped, the apex obtuse; pollinia 2, obovate-subglobose, sulcate; stipe spathulate, hyaline, recurved below and between the curved apex of the bifid rostellum; stigma suborbicular, concaved. Capsules cylindric-fusiform. Epiphytic monopodial orchids with short stems bearing 4–5 oblong-elliptic leaves near the apex and many sheathing remains of fallen leaves mixed with numerous thick aerial roots. Panicles with short peduncles and 2–3 slender branches, each bearing 20–25 flowers opening simultaneously. One species collected by Charles Ford in 1883, from Loh-fau Shan, flowered in the Botanic Gardens of Hong Kong in 1884, *O. eublepharon* Hance. Many species of orchids formerly described from Loh-fau Shan have been recently found from Tai Mo Shan in the New Territories of Hong Kong. *Ornithochilus eublepharon* has not been found yet. There may be a possibility of locating it in Tai Mo Shan or Lantau Island.

Ornithochilus eublepharon Hance has been subjected to two nomenclatural changes. It was first made a synonym of *O. fuscus* Wall. ex Lindl. and then transferred to *Sarchochilus* and it became synonymous to *S. difformis* (Wall. ex Lindl.) Tang & Wang. It is recognized here for the following reasons. Hance described *O. eublepharon* on the basis of Ford's collection from Loh-fau Shan. The isotype of the species shows that it has paniculate inflorescences, ovate lateral sepals, oblong petals and strongly bilobed fimbriate midlobe of the lip. According to Hooker's description, the species from the Himalayan Region, *O. fuscus* Wall. ex Lindl., has racemose inflorescences, oblique-obovate sepals, linear petals and entire fimbriate midlobe of the lip. Both geographical separation and morphological features disprove that *O. eublepharon* and *O. fuscus* are conspecific. *Sarcochilus* is typified by *S. fulcatus* R. Br., a species from Australia and with 4 pollinia. *Ornithochilus eublepharon* has 2 pollinia.

Figure 68. *Robiquetia succisa*: A. The habit sketch of a fruiting branch showing the long aerial roots, and the monopodial stem with distichously arranged leaves irregularly erose, emerginate and apiculate at the apex. B. The habit sketch of a flowering branch showing a flowering panicle with some flowers and buds. C. The front view of a flower showing the sepals, petals, lip, column, and the anther rostrate at the apex. D. The lateral view of a flower showing the sepals, petals, and the lip with a thick spur swollen at the tip. E. The same with half of the odd sepals, a lateral sepal, a petal, and a portion of the lip removed, showing a horizontal flap on the back at the entrance to the spur, 2 longitudinal ridges on the disc, a wing-like truncate lateral lobe, and the tongue-like apical lobe of the lip. Also shown are the short stout column with the anterior portion extended forward into two arms, and the anther with slender long apical projection. F. The front view of the spur showing the swollen apex. G. The front view of the pollinia on a short stipe, and the curved caudicle attached to the viscidium at the apex. H. The lateral view of the same.

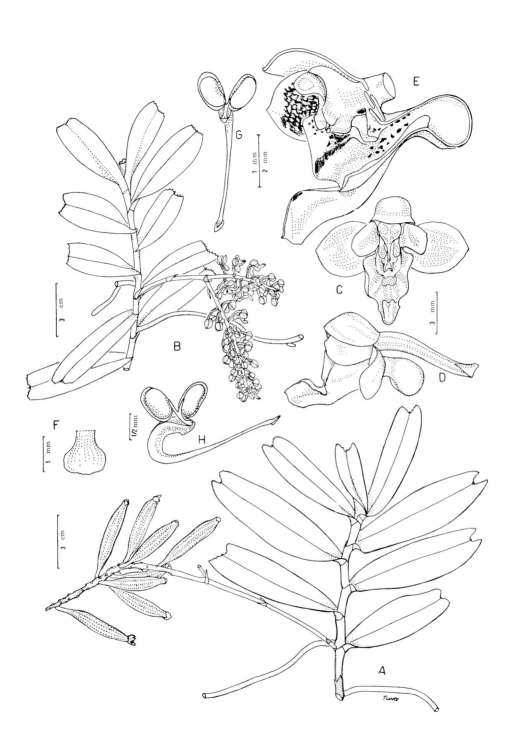

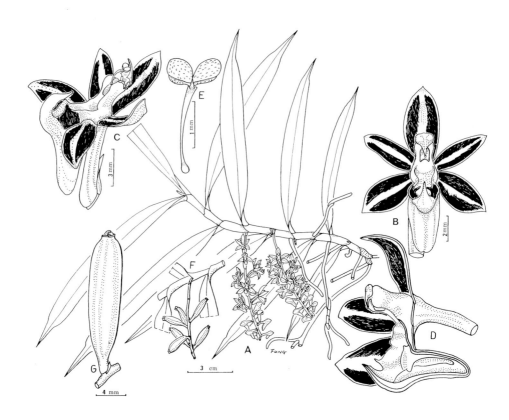

FIGURE 69. *Cleisostoma fordii*: A. The habit sketch of a flowering plant showing the mono-podial stem, the distichous leaves, and the position of the inflorscences. B. The front view of a flower showing the sepals, petals, saccate lip with horn-like processes on the triangular lateral lobes and the pointed midlobe, the column with rostrate anther cap, and the stigma. C. The lateral view of a flower showing the sepals, petals, saccate lip, the column with terminal anther cap slightly tipped backward, and the paired pollinia on a slender stipe. D. The lateral view of a flower with one-half of the odd sepal, a lateral sepal, a petal, and one-half of the lip removed, showing the callus on the back at the orifice of the spur fitted to the front keel running down from the disc of the lip to the bottom of the spur. E. Four pollinia in 2 pairs on an elongated stipe. F. A fruiting branch showing six capsules. G. An oblong-cylindric capsule with persistent column.

FIGURE 70. *Ornithochilus eublepharon*: A. The habit sketch of a flowering plant showing the monopodial growth with numerous aerial roots, distichous leaves, and a panicle with 2 racemes. B. The laterial view of a flower with a lateral sepal omitted, showing a small bract, an ovate lateral sepal, the oblong petals, the calcarate lip attached to the column-foot, the column, and the anther. C. The lip as viewed from the left and top angle, showing the narrowed base, the erect and slightly incurved lateral lobes, the obcordate-reniform unguiculate midlobe with a basal glandular flap over the orifice of the spur and fimbriate margin. D. The abaxial view of the midlobe showing the bilobed apex and the fringed margin. E. The apical portion of the column with the anther slightly pushed back, showing the broad stipe over the rostellum and recurved at the apex. F. Two pollinia on a broad hyaline stipe recurved at the apex of the rostellum (Illustration prepared from Ford's collection, Hong Kong Herbarium).

118

FIGURE 71. Flower color of four spontaneous species: A. *Paphiopedilum purpuratum*. B. *Brachycorythis galeandra*. C. *Pecteilis susannae*. D. *Habenaria linguella*. Photo: James Leung.

FIGURE 72. Flower color of four spontaneous species: A. *Peristylus goodyeroides*. B. *Cheirostylis chinensis*. C. *Goodyera procera*. D. *Zeuxine gracilis*. Photo: James Leung.

FIGURE 73. Flower color of five spontaneous species: A. *Dendrobium loddigesii*. B. *Dendrobium hercoglossum*. C. *Arundina chinensis*. D. *Liparis nervosa*. E. *Bulbophyllum youngsayeanum*. Photo: James Leung.

FIGURE 74.   Flower color of five spontaneous species: A. *Eria rosea*.   B. *Geodorum densiflorum*.
C. *Thrixspermum centipeda*.   D. *Diploprora championii*.   E. *Cleisostoma teres*.   Photo: James Leung.

III
Composition and
Phytogeographic Significance

In 1971 an illustrated account of the composition and distribution of orchids in China was published.[1] In this article 17 genera of Hong Kong orchids were represented in 14 distributional maps. Additional information for 22 other genera was included in a chart and its explanation. With this as a basis, a systematic summary of the available data on the Orchidaceae of Hong Kong, an analysis of the composition of the family in the vegetation of the area, and an interpretation of the phytogeographic significance of these orchids are prepared.

Systemic Summary

For the understanding of the composition and the phytogeographic significance of the Orchidaceae in Hong Kong, a systematic arrangement of the genera of the area, together with information on the world features, and the number of species of the genera in China is given in Table I. The system adopted is the generally recognized one which was proposed by R. Schlechter[2] in 1926, and used by orchidologists in China.

In the table, in order to distinguish the native and the exotic elements, the subtribes and genera which occur in Hong Kong only in cultivation are in capitals. In two genera, *Paphiopedilum* and *Dendrobium*, where both native and introduced species are present in Hong Kong, two numbers are given, with the first number indicating the native species and the one after the plus sign (+) representing the exotic species. In the column on world features, the term "Southeast Asia" refers to the area which includes the subtropical and tropical area of South China, thence westward to northeastern India and Burma, and eastward to Taiwan, the Philippines, and southward to Indo-China, Thailand, Malaysia, and Indonesia.

---

[1] Hu, S. Y. (1971) The Orchidaceae of China II. Quart. Journ. Taiwan Mus. **24**:181–255.
[2] Schlechter, R. (1926) Das System der Orchidaceen, Notizbl. Bot. Gart. Mus. Berlin-Dahlem **9**:563–591.

Table I. A Systematic Summary of the Genera of Orchidaceae in Hong Kong

SYSTEMATIC ARRANGEMENT *Subfamilies, tribes, subtribes, and genera*	WORLD FEATURES *Total no. of species*	*General distribution*	NUMBER OF SPECIES *in China*	*in Hong Kong*
Subfamily **Cypripedioideae**—Tribe **A. Cypripedieae**				
Cypripediinae				
Paphiopedilum	50	Southeast Asia, chiefly Burma, Philippines.	7	1 + 3
Subfamily **Orchidoideae**—Tribe **B. Orchideae**				
Orchidinae				
Brachycorythis	50	Africa chiefly.	2	1
Platanthera	200	East and Southeast Asia.	57	3
Habenariinae				
Pecteilis	6	Southeast Asia.	2	1
Peristylus	60	Southeast Asia to Africa.	11	5
Habenaria	600	Widespread, chiefly North temperate regions.	108	6
Disperidinae				
Disperis	100	Chiefly Africa, Madagascar, Pacific Is., 1 Taiwan.	1	1
Tribe **C. Neottieae**				
Cryptostylidinae				
Cryptostylis	20	Southeast Asia to Australia.	3	1
Cephalantherinae				
Aphyllorchis	20	Southeast Asia to New Guinea.	5	1
BLETILLINAE				
BLETILLA	10	East Asia, chiefly China.	9	1
Nerviliinae				
Nervilia	80	Southeast Asia to Africa.	6	1
Spiranthinae				
Spiranthes	25	Widespreading, chiefly tropical America.	4	2
Manniellinae				
Manniella	3	Africa, S. Am., Hong Kong.	0	1
Erythrodiinae				
Goodyera	165	East and Southeast Asia to Papuasia; N. Am. and W. Indies.	39	4
Ludisia (Haemaria)	6	Southeast Asia.	1	1
Cheirostylis	22	Southeast Asia to New Guinea.	9	4
Zeuxine	76	Southeast Asia to New Guinea.	14	4
Hetaeria	20	Himalayan Reg., Southeast Asia to Sri Lanka, eastward to New Guinea and New Caledonia.	4	2
Anoectochilus	25	Eastern Himalayan Reg. to South China and Malaysia.	4	1
Vrydagzynea	40	Hong Kong, Taiwan southward to Malaysia and eastward to Solomon Is. and New Guinea.	2	1

TABLE I. (cont'd)

SYSTEMATIC ARRANGEMENT Subfamilies, tribes, subtribes, and genera	WORLD FEATURES Total no. of species	General distribution	NUMBER OF SPECIES in China	in Hong Kong
Tropidiinae				
Tropidia	22	Hong Kong, Taiwan southward to Malaysia, Sri Lanka, India.	5	1
Tribe D. Epidendreae—Series **Acranthae**				
Liparidinae				
Malaxis	300	Pan tropic.	19	4
Liparis	250	Old World Tropics.	57	7
Collabiinae				
Mischobulbum	7	Hong Kong, Taiwan, southward to Malaysia.	1	1
Tainia	25	Eastern Himalayan Reg. eastward to New Guinea.	13	1
Ania	11	Southeast Asia.	4	2
Nephelaphyllum	12	Kwangtung, Hong Kong, southward to Malaysia.	2	1
Coelogyniinae				
Coelogyne	200	Eastern Himalayan Reg. to Southeast Asia to Borneo.	29	2
Pholidota	55	Eastern Himalayan Reg., Southeast Asia to Papuasia.	14	2
Thuniinae				
Arundina	20	Eastern Himalayan Reg. to Malaysia.	2	1
EPIDENDRINAE				
EPIDENDRUM	400	Tropical America.	22	1
CATTLEYA	60	Tropical America.	37	2
SOPHRONITIS	6	Brazil.	1	1
Dendrobiinae				
Dendrobium	900	Southeast Asia, chiefly Philippines and Borneo.	79	3 + 4
Eria	350	Southeast Asia eastward to Papuasia.	28	5
Podochilinae				
Appendicula	100	Eastern Himalayan Reg. eastward to Southeast Asia and Papuasia.	7	1
Tribe D. Epidendreae—Series **Pleuranthae**				
Phaiinae				
Calanthe	120	Southeast Asia eastward to New Caledonia and southwestward to Africa.	67	4
Cephalantheropsis	6	Southeast Asia, chiefly the Philippines.	1	1
Phaius	50	Southeast Asia eastward to New Guinea and westward to India and Madagascar.	8	1 + 1 variety
Pachystoma	12	Southeast Asia eastward to New Caledonia.	4	1
Acanthephippium	15	Southeast Asia to Fiji.	4	1
Spathoglottis	46	Southeast Asia to New Caledonia.	3	2

TABLE I. (*cont'd*)

SYSTEMATIC ARRANGEMENT Subfamilies, tribes, subtribes, and genera	WORLD FEATURES		NUMBER OF SPECIES	
	Total no. of species	*General distribution*	*in China*	*in Hong Kong*
Bulbophyllinae				
Bulbophyllum	900	Southeast Asia eastward to New Guinea and southwestward to tropical Africa.	34	5
Cirrhopetalum	70	Southeast Asia, chiefly South China, Burma, Thailand, Philippines, and Malaysia.	25	5
Cyrtopodiinae				
Geodorum	16	Eastern Himalayan Reg., Southeast Asia to Polynesia.	6	1
Eulophia	200	Pantropic, largely Africa.	17	3
Cymbidiinae				
Cymbidium	300	Tropics of both hemispheres, largely Asia.	49	5 + 1 and 1 variety
Thelasiinae				
Thelasis	20	Eastern Himal. Reg. to Malaysia.	3	1
LYCASTIINAE				
LYCASTE	45	Tropical America.	1	1
ONCIDIINAE				
ODONTOGLOSSUM	200	Tropical America.	2	2
MILTONIA	25	Tropical America.	2	1
ONCIDIUM	350	Tropical America.	12	1
Sarcanthinae				
PHALAENOPSIS	35	Tropical Asia, Sri Lanka to Philippines and New Guinea.	3	3 + 1 cultivar
Diploprora	6	N. Thailand, Hong Kong, Taiwan.	3	1
Thrixspermum	100	Southeast Asia, chiefly Philippines.	10	1
ARACHNIS	7	Southeast Asia, chiefly Philippines.	2	2
Cleisostoma	100	Eastern Himalayan Reg., Southeast Asia.	20	3
Robiquetia	10	Tropical Asia.	6	1
Ornithochilus	6	Eastern Himalayan Reg., eastward to New Guinea and Australia.	3	1 (?)
Gastrochilus	20	Eastern Himalayan Reg. to Malaysia.	10	1
RENANTHERA	12	Southeast Asia to Solomon Is.	3	1
VANDA	60	Southeast Asia to Australia.	11	3
Acampe	15	Southeast Asia to tropical Africa.	1	1
MICROCOELIA	30	Tropical Africa.	1	1

Table I shows three outstanding features of the Orchidaceae of Hong Kong. These are: (1) a close affinity with the orchids of China at the generic level; (2) morphological diversity at the supergeneric level; and (3) small size of species.

1. *Generic Affinity with the Orchids of China*: Geologically, geographically, and floristically Hong Kong is inseparable from South China. The composition of the Orchidaceae conforms with this fact. With the exception of *Manniella*, all the native genera of orchids in Hong Kong are represented in China. The case of *Manniella* is unique. The species in this genus is deciduous. It requires a period of dormancy in the cool dry monsoon season. For three or four months of the year, it disappears from the surface of the earth and stays dormant underground. As it resumes growth in the spring, it flowers without leaves (Fig. 15A). The flowers are inconspicuous, few, small, hardly open and ephemeral. When the vegetative shoot appears, it is acaulescent, plantain-like, consisting of a rosette of rather broad elliptic leaves which give no impression of an orchid. With all these unusual characteristics, perhaps species of the genus in the subtropical and tropical coastal areas and islands of China have escaped the notice of former collectors. It is probable that the same condition exists elsewhere, and consequently only a few species are known from the tropics of both hemispheres. The discovery of the genus in China may be achieved by more extensive and intensive explorations of the future.

The genera with large number of species in China also have higher number of species in Hong Kong. *Bulbophyllum, Calanthe, Cirrhopetalum, Cleisostoma, Coelogyne, Cymbidium, Dendrobium, Eulophia, Goodyera, Habenaria, Liparis, Malaxis, Peristylus, Pholidota, Platanthera,* and *Zeuxine* are the examples.

Although there is a general agreement between the genera of orchids common to China and Hong Kong, the Orchidaceae of Hong Kong does have unique characteristics. In Hong Kong, the tropical and subtropical orchids show greater development than the temperate types. For example, in China, the genus with the highest number of species is *Habenaria*, and that of Hong Kong is *Liparis*. *Habenaria* is a deciduous genus with worldwide distribution, chiefly in the North Temperate zone and at higher altitudes in the mountains of the subtropical and the tropical regions where temperate elements prevail. The aerial portion of the species disappears in winter. A new leafy and flowering shoot develops from a subterranean tuber in the next growing season. In Hong Kong, species of this genus grow in hilltop meadows on open slopes where they stay dormant in the dry monsoon season. On the other hand, *Liparis* is an evergreen genus of the Old World Tropics. Its new shoot emerges from the base of a pseudobulb. China, with its extensive temperate flora in the North and on the mountains in the West and South, supports an extremely rich temperate flora, while in Hong Kong the flora is better represented by subtropical and tropical elements. This fact is particularly well expressed in *Bulbophyllum, Cheirostylis, Cirrhopetalum, Cleisostoma, Eulophia, Liparis, Malaxis, Peristylus,* and *Zeuxine*. These genera have proportionately larger number of species in Hong Kong.

2. *Morphological Diversity*: Morphological diversity is the most prominent feature of the Orchidaceae in Hong Kong. With the exclusion of the 14 introduced genera which are inconsequential in the analysis of the composition of the natural vegetation of an area, the 50 native genera are in 23 subtribes, four tribes and two subfamilies of

the Schlechterian System. This is to say that among the native species of orchids in an area of 404 square miles of Hong Kong, there are representatives of all the subfamilies, tribes, series, 59% of the subtribes, and 33% of the genera of the Orchidaceae in China. This phenomenon indicates a fantastic complexity in the composition of the Orchidaceae of Hong Kong. The morphological characteristics of all the genera of these orchids are described with illustrations in chapter II of this work. It is fitting to add here that in August and September of 1975, a member of the Native Orchid Group added two new records to the formerly known list of the genera and species of orchids of the area. One of the taxa appears to be a parasite, for it is aphyllous and without fleshy rhizomes or tubers. These findings indicate that our knowledge of the morphological diversity of Hong Kong Orchidaceae is not exhausted. Future explorers may find additional material with distinctive characters not yet known to exist in this area.

Morphological diversity of the Orchidaceae in Hong Kong is a reflection of the physical condition of the area which provides favorable ecological niches for the growth of many different forms of orchids. In the analysis of the composition of the Orchidaceae of China, the various patterns of distribution reveal that within the country, there are orchid-rich and orchid-poor regions. The maps show that Hong Kong is in the center of the orchid-rich belt of China which extends from Yunnan and its adjacent provinces eastward to Taiwan. The geological history and the heterogenous topography of Hong Kong provide diversified ecological conditions which support various life forms. The examples are: (1) the hilltop meadows for the deciduous genera such as *Habenaria* and *Platanthera*; (2) the exposed boulders and arid cliffs for the xerophytic saxicolous genera such as *Cleisostoma*, *Coelogyne*, and *Eria*; (3) the less accessible wooded ravines and steep gorges with running streams for the creeping genera such as *Goodyera* and *Anoectochilus* on the wet forest floor, the shade-loving epiphytic genera such as *Bulbophyllum*, *Acampe*, and *Liparis* on the damp rocks or tree trunks and *Diploprora* and *Thrixspermum* on the partially shaded perpendicular cliffs, and the thickets of lianas and small trees by the streams for the saprophytic or parasitic genera such as *Aphyllorchis* and *Eulophia* (*E. yushuiana*), or for the symbiotic genus *Cheirostylis* on moss-covered boulders; and (4) the windy scrubby slopes for the sun-loving genera such as *Pachystoma*, *Spathoglottis*, and *Eulophia* (*E. flava* and *E. sinensis*).

Geological background and heterogeneous topography are important for they provide a diversified ecological setting for various form of orchids. However, it is the favorable climatic condition of Hong Kong which exerts the controlling factor and accounts for the rich orchid flora of the area. A normally frost-free climate, with an alternation of a hot humid wet season and a pleasant mild dry one the latter half of which is cloudy and cool, with frequent light rains, provides the optimum condition for the growth of orchids. Water is extremely important for the survival of orchids. In Hong Kong, water is adequate for the existence of the native orchids. In the wet season, there is a plantiful supply of water in the soil for the terrestrial species and high humidity in the air for the epiphytic and creeping species. In the winter monsoon season, Hong Kong receives an average of one-seventh of the mean value of the annual precipitation (2 168.8 mm. = 85.39 inches).[3] Even in the pleasant dry period,

---

[3] Dwyer, D. J. (1969) Geography and climate. Hong Kong, Report for the Year 1968, pp. 246–55.

there is often a shower at dawn or in the early morning. The distribution of the rainfall of Hong Kong is uneven, with Lantau and Tai Mo Shan receiving heavier rains, and the peaks and higher altitudes being wetter than the low areas. The seepage areas of the hillsides and the wooded ravines of Hong Kong is never too dry or too cold for the growth of tropical and subtropical orchids, while the hilltops and the exposed scrubby ridges and slopes are dry and cold enough to support the species of the deciduous genera which are more widespread in the temperate regions.

It is appropriate therefore to conclude that the combined effect of the favorable natural environmental factors is responsible for the rich assemblage of morphologically diversified orchids in Hong Kong.

It is worthy of note that the morphological characters of a majority of the cultivated orchids are different from those of the native species. Approximately 60% of the exotic species in Hong Kong are native of tropical America. They are introduced for the colorful showy flowers. Another 30% of them are introduced from the Philippines, Thailand, Indo-China, or Malaysia. *Bletilla striata* is from temperate China, and a few species of *Cymbidium* and *Dendrobium*, and *Renanthera coccinea* are Chinese favorites of the warmer region since time immemorial. With few exceptions, *Phaius tankervilliae* and *Paphiopedilum purpuratum* are the native species that may occur in private collections, and these species are becoming very rare in nature.

3. *Small Species and Poor Populations*:   The Orchidaceae of Hong Kong is characterized by having low numbers of species and small populations in most genera. Approximately 60% of the genera have one species each. In addition, 12% of the genera have 3 species each, and 10% 2 species each. The largest genus of orchids in Hong Kong is *Liparis* which has 7 species, and the next large genus is *Habenaria* which has 6 species. The genera with 5 species are *Bulbophyllum, Cirrhopetalum, Cymbidium, Eria,* and *Peristylus*. Those with four species are *Calanthe, Cheirostylis, Goodyera, Malaxis,* and *Zeuxine*.

The Orchidaceae in Hong Kong is characterized not only by the low number of species, it is also marked by the paucity in the populations. Orchidaceae is the third largest family in the flora of Hong Kong. The other large families of the area are Leguminosae, Gramineae, and Compositae. Unlike members of these families, the species of Orchidaceae are inconspicuous in the landscape. With the exception of some amateurs or economically minded village people who go out to look for orchids in the underdeveloped areas, few residents of Hong Kong have seen native orchids in their natural habitats. There are three causes for the inconspicuousness and paucity in population of native orchids in Hong Kong. Two of these are due to the nature of the orchids. These are (1) the small size and the lack of showy flowers, and (2) the preference for undisturbed habitats. The third cause is man-made.

Approximately 50% of the species of native orchids of Hong Kong have small greenish yellow, grayish, or white flowers of 5 mm. or less across. They generally grow among grasses, sedges, or mosses. It takes the trained eyes of keen observers to see these species even at anthesis. For example, there are three colonies comprising dozens to hundreds of individuals of *Eulophia sinensis, Spiranthes hongkongensis,* and *Zeuxine strateumatica* in the campus of The Chinese University. An estimate of less then one-hundredth of one percent of the people of the University community have seen any of these species because of the small size of their flowers.

The native orchids of Hong Kong prefer undisturbed habitats. With the exception of the above mentioned three species, most of Hong Kong orchids occur in wooded ravines or exposed hilltops and slopes. Approximately 66% of the native genera occur in the shade of forest floors or on the wet rocks of wooded ravines. About 22% of the genera grow in meadows of arid hilltops, in depressions or swampy areas of open hillsides, or among grasses near abandoned farms. Some of the grassland species produce attractive flowers of considerable size. *Arundina chinensis, Habenaria dentata, H. linguella,* and *Eulophia flava* are some examples. The most abundant species of the meadows is *Spathoglottis pubescens*. During the flowering season, this species may display a spectacular sight on certain slopes of Lantau Island, or in some areas on the northern flanks of Pat Sin Range (八仙嶺). About 10% of the native orchid genera are saxicolous. They grow attached to barren rocks subjected to diurnal changes of temperature and to strong winds. People who stay in the urban areas normally do not see the native orchids.

The man-made causes of the paucity of native orchids are twofold; repeated fires of the hillsides, and ruthless collection and uprooting. The periodical burning of the hillsides is associated directly with the semiannual mountain day for ancestor worship, and with the building of camp-fires. Fires carelessly left alight may lead to continuous burning of the hillsides for two or three days, destroying the plants and animals, altering the physical and chemical properties of the soil, and consequently changing the vegetation from forests to grassland.

Ruthless collection and uprooting is due to the activity of the suppliers of herbalists and aquarium owners. Eight species of native orchids are known to be used for medicine in Hong Kong. These are *Arundina chinensis, Bulbophyllum radiatum, Dendrobium hercoglossum, D. loddigesii, Ludisia discolor, Nervilia fordii, Pholidota chinensis,* and *Spiranthes sinensis*. All the species are uprooted and the entire plants used in medicine. With 90% of the population in Hong Kong taking herbal medicine in one way or the other, there will be fewer and fewer native orchids as well as other species of plants available to the herbalists and the people. For this reason, the highly esteemed medicinal orchid species such as *Ludisia discolor* and *Nervilia fordii* are on the verge of extinction.

Recently in Hong Kong, there has been large scale collection of non-medicinal orchids for the decoration of aquaria. It has been observed that *Goodyera procera, Liparis longipes, Pholidota chinensis,* and *Paphiopedilum purpuratum* are available at the aquarium supply market on the sidewalk of Mong Kok Station in Kowloon for sale in the early morning. *Goodyera procera* grows on the mixture of sand and plant debris collected in the depressions or old potholes in the course of the upper level of the streams. Its strong roots are able to hold fast when the current becomes strong in the rainy season, and to accumulate more sand and organic matter in slow current. *Liparis longipes* and *Pholidota chinensis* grow on wet rocks in shady areas. Although these species grow in association of streams, their leaves are never submerged in water for any length of time. As their size and form are good for the decoration of aquaria, people in the urban area buy them gladly. When they die in submersion, the aquarium owners blame for their own inadequate technique in caring for them. Thus there is a continuous demand for these species. *Paphiopedilum purpuratum* is doomed to the same destiny. The natural supply of the species of *Goodyera, Liparis,* and *Pholidota* is more abundant than that of *P. purpuratum*. The distribution of *P. purpuratum* is extremely limited, and this species is now on the verge of extinction due to unwise use and ruthless uprooting. Other species of orchids that are extremely rare or have disappeared from Hong Kong are *Aphyllorchis montana, Acanthephippium sinense, Anoectochilus yungianus, Thelasis hongkongensis,* and *Tropidia hongkongensis.*

4. *Alphabetic List of the Species*: For the dual purposes of providing a reference to the component species of Orchidaceae in the flora of Hong Kong, and of enabling the readers to locate readily the genera and species covered in the descriptions and keys of this work, this alphabetic list is prepared. It covers both the spontaneous and the cultivated species. The number in the parentheses after each name represents the position of the genus included both in the keys and in the sequence of the descriptions. Various editions of Check List of Hong Kong Plants have been prepared by the Hong Kong Herbarium and issued between 1962 and 1974. The names which appear in the fourth edition (1974) of the Check List but are excluded from this work are in *italics*, each followed by a cross reference or an explanation. Names with asterisks (*) represent exotic species or cultivars which are known in Hong Kong in cultivation only.

The names of authors of the species are given in full wherever space permits and whenever information is available. This is done for the convenience of readers who may like to know the names of the botanists who made the original descriptions of the species or who transferred the names to other genera of orchids that now grow in Hong Kong.

An Alphabetic List of the Species of Orchids Records from Hong Kong

Acampe multiflora (Lindley) Lindley (59)
Acampe rigida (Buch.-Ham. ex J. E. Sm.) Hunt—a Himalayan species.
Acanthephippium sinense Rolfe (44)
Ania hongkongensis (Rolfe) Tang & Wang (29)
Ania ruybarrettoi Hu & Barretto (29)
Anoectochilus yungianus S. Y. Hu (13)
Aphyllorchis montana Reichenbach f. (20)
Appendicula bifaria Lindley ex Bentham (22)
*Arachnis flos-aeris** (L.) Reichenbach f. (54)
*Arachnis maingayi** (Hooker f.) Schlechter (54)
Arundina chinensis Blume (24)
Arundina chinensis var. **major** S. Y. Hu (24)
Arundina graminifolia (D. Don) Hochreutiner—a species of northern India, 1–2 m. tall.
*Bletilla striata** (Thunberg) Reichenbach f. (32)
Brachycorythis galeandra (Reichenbach f.) Summerhayes (3)
Bulbophyllum ambrosia (Hance) Schlechter (41)
Bulbophyllum levinei Schlechter (41)
Bulbophyllum odoratissimum Lindley (41)
Bulbophyllum radiatum Lindley (41)
Bulbophyllum watsonianum Reichenbach f.=B. ambrosia (Hance) Schlechter.
Bulbophyllum youngsayeanum Hu & Barretto (41)
Calanthe gracilis Lindley=Cephalantheropsis gracilis (Lindley) S. Y. Hu
Calanthe masuca Lindley ex Wallich—Hong Kong population belongs to the following variety.
Calanthe masuca var. **sinensis** Rendle (46)
Calanthe patsinensis S. Y. Hu (46)
Calanthe striata (Swartz) R. Brown (46)
Calanthe triplicata (Willemet) Ames (46)
Calanthe veratrifolia R. Brown—Hong Kong material=C. triplicata (Willemet) Ames.

*Cattleya lueddemanniana Reichenbach f. (37)
*Cattleya mendelii Backhouse (37)
 Cephalantheropsis gracilis (Lindley) S. Y. Hu (21)
 Cheirostylis chinensis Rolfe (12)
 Cheirostylis clibborndyeri Hu & Barretto (12)
 Cheirostylis flabellata (A. Richard) Wight—an Indian species, not in Hong Kong.
 Cheirostylis jamesleungii Hu & Barretto (12)
 Cheirostylis monteiroi Hu & Barretto (12)
 Cirrhopetalum bicolor (Lindl.) Rolfe (40)
 Cirrhopetalum delitescens (Hance) Rolfe (40)
 Cirrhopetalum miniatum Rolfe (40)
 Cirrhopetalum tigridum (Hance) Rolfe (40)
 Cirrhopetalum tseanum Hu & Barretto (40)
 Cleisostoma fordii Hance (63)
 Cleisostoma paniculata (Hance) Garay (63)
 Cleisostoma teres Garay (63)
 Cleisostoma virginale Hance = Robiquetia succisa (Lindley) Seidenfaden & Garay
 Coelogyne fimbriata Lindley (39)
 Coelogyne leungiana S. Y. Hu (39)
 Cryptostylis arachnites (Blume) Hasskarl (9)
 Cymbidium ensifolium Swartz (49)
 Cymbidium lancifolium Hooker (49)
 Cymbidium maclehoseae S. Y. Hu (49)
 Cymbidium pendulum (Roxburgh) Swartz (49)
 Cymbidium sinense (Andrews) Willdenow (49)
 Cymbidium sinense var. albo-jucundissimum (Hayata) Fukuyama (49)
 Cymbidium xiphiifolium Lindley (49)
 Dendrobium acinaciforme Roxburgh (23)
 Dendrobium hercoglossum Reichenbach f. (23)
*Dendrobium fimbriatum Hooker (23)
*Dendrobium linawianum Reichenbach f. (23)
 Dendrobium loddigesii Rolfe (23)
*Dendrobium moniliforme (L.) Swartz (23)
*Dendrobium nobile Lindley (23)
 Diploprora championii (Lindley) Hooker f. (56)
 Disperis lantauensis S. Y. Hu (7)
*Epidendrum ibaguense Humboldt, Bonpland & Kunth (25)
 Eria corneri Reichenbach f. (42)
 Eria coronaria (Lindley) Reichenbach f.—a species from northern India, material from Hong
    Kong misidentified.
 Eria flava Lindley (42)
 Eria rosea Lindley (42)
 Eria sinica Lindley (42)
 Eulophia flava Hooker f. (31)
 Eulophia sinensis Miquel (31)
 Eulophia yushuiana S. Y. Hu (31)
 Gastrochilus holttumianus Hu & Barretto (61)
 Geodorum densiflorum (Lamarck) Schlechter (48)
 Geodorum dilatatum R. Brown = G. densiflorum (Lam.) Schlechter
 Geodorum semicristatum Lindley—former records not yet substantiated.
 Glossula calcarata Rolfe = Peristylus calcaratus (Rolfe) S. Y. Hu
 Goodyera cordata (Lindley) Bentham ex Hooker f. (15)
 Goodyera foliosa (Lindley) Bentham ex Hooker f. (15)
 Goodyera procera (Ker-Gawler) Hooker (15)
 Goodyera youngsayei Hu & Barretto (15)
 Habenaria ciliolaris Kränzlin (5)
 Habenaria dentata (Swartz) Schlechter (5)

Habenaria leptoloba Bentham (5)

Habenaria linguella Lindley (5)

Habenaria mandarinorum—not a valid name, see *Platanthera mandarinorum* Reichenbach f.

Habenaria reniformis (D. Don) Hooker f. (5)

Habenaria rhodocheila Hance (5)

Habenaria stenostachya (Lindl.) Benth. = Peristylus densus (Lindl.) Santapau & Kapadia

Habenaria tentaculata Reichb. f. = Peristylus tentaculatus (Lindl.) J. J. Smith

Haemaria discolor (Ker-Gawl.) Lindl. = Ludisia discolor (Ker-Gawler) A. Richard

Hetaeria cristata Blume (17)

Hetaeria nitida Ridley (17)

Liparis chloroxantha Hance (35)

Liparis longipes Lindley (35)

Liparis macrantha Rolfe (35)

Liparis nervosa (Thunberg) Lindley (35)

Liparis odorata (Willdenow) Lindley (35)

Liparis plicata Franchet & Savatier (35)

Liparis ruybarrettoi Hu & Barretto (35)

Ludisia discolor (Ker-Gawler) A. Richard (14)

*****Lycaste skinneri** Lindley (47)

Malaxis acuminata D. Don var. **biloba** (Lindley) Hunt (36)

Malaxis allanii Hu & Barretto (36)

Malaxis calophylla (Reichb. f.) O. Kuntze—former record not substantiated.

Malaxis latifolia J. E. Smith (36)

Malaxis parvissima Hu & Barretto (36)

Manniella hongkongensis Hu & Barretto (11)

*****Microcoelia guyoniana** (Reichenboch f.) Summerhayes (58)

*****Miltonia** "Belt Field" (52)

Mischobulbum cordifolium (Hooker f.) Schlechter (27)

Nephelaphyllum cristatum Rolfe (26)

Nervilia fordii (Hance) Schlechter (8)

*****Odontoglossum grande** Lindley (51)

*****Odontoglossum pendulum** Bateman (51)

*****Oncidium varicosum** Lindley (50)

*****Ornithochilus eublepharon** Hance (64)

Ornithochilus fuscus Wallich ex Hooker f.—a species of the Himalayan Region.

Pachystoma chinense Reichenbach f. (30)

*****Paphiopedilum appletonianum** Rolfe (1)

*****Paphiopedilum exul** Rolfe (1)

*****Paphiopedilum haynaldianum** (Reichenbach f.) Pfitzer (1)

Paphiopedilum purpuratum Pfitzer (1)

Pecteilis susannae (L.) Rafinesque (2)

Peristylus calcaratus (Rolfe) S. Y. Hu (6)

Peristylus chloranthus Lindley = P. spiranthes (Schauer) S. Y. Hu

Peristylus densus (Lindley) Santapau & Kapadia (6)

Peristylus goodyeroides (D. Don) Lindley (6)

Peristylus spiranthes (Schauer) S. Y. Hu (6)

Peristylus stenostachyus (Lindl.) Kränzlin = P. densus (Lindl.) Sant. & Kap.

Peristylus tentaculatus (Lindley) J. J. Smith (6)

Phaius grandifolius Loureiro = P. tankervilliae (Banks ex l'Heritier) Blume

Phaius longipes (Hook. f.) Holttum = Cephalantheropsis gracilis (Lindley) S. Y. Hu

Phaius tankervilliae (Banks ex l'Heritier) Blume (45)—Perianth brown within.

Phaius tankervilliae f. **veronicae** Hu & Barretto (45)—Perianth lemon-yellow within.

*****Phalaenopsis amabilis** (L.) Blume (57)

*****Phalaenopsis equestris** Reichenbach f. (57)

*****Phalaenopsis schilleriana** Reichenbach f. (57)

*****Phalaenopsis** "Star of Leyte" (57)

Pholidota cantonensis Rolfe (38)

135

Pholidota chinensis Lindley (38)
Platanthera angustata (Blume) Lindley (4)
Platanthera mandarinorum Reichenbach f. (4)
Platanthera minor (Miq.) Reichenbach f.
Pomatocalpa virginale (Hance) J. J. Smith=Robiquetia succisa (Lindl.) Seid. & Garay
*****Renanthera coccinea** Loureiro (55)
Robiquetia succisa (Lindey) Seidenfaden & Garay (62)
Sarcanthus fordii (Hance) Rolfe=Cleisostoma fordii Hance
Sarcanthus hongkongensis Rolfe—Although this species is named for Hong Kong, there is no
material evidence that it occurs in the area. It was described on the basis of a plant sent to
the Royal Botanic Gardens at Kew by Charles Ford of the Hong Kong Government in the
late 1880s. The plant flowered in Kew in 1897. Ford was an ardent botanical explorer in
South China as well as an able administrator in Hong Kong. As indicated by his specimens,
he collected from northern Kwangtung to Hainan Island. In Hong Kong Herbarium there
is no specimen of *S. hongkongensis* collected by Ford, but there is one from Teng Woo Shan
of northern Kwangtung, *C. Wang 34744*. This collection fits with Rolfe's description of the
species.
Sarcanthus teretifolius Lindley=Cleisostoma teres Garay
*****Sophronitis grandiflora** Lindley (43)
Spathoglottis fortunei Lindley (33)
*****Spathoglottis plicata** Blume (33)
Spathoglottis pubescens Lindley (33)
Spiranthes hongkongensis Hu & Barretto (10)
Spiranthes sinensis (Persoon) Ames (10)
Tainia dunnii Rolfe (28)
Tainia hongkongensis Rolfe=Ania hongkongensis (Rolfe) Tang & Wang
Tainia viridifusca (Hook.) Bentham=Ania viridifusca (Hook.) Tang & Wang, a name for a
Himalayan species, material from Hong Kong misidentified.
Thelasis hongkongensis Rolfe (34)
Thrixspermum centipeda Loureiro (53)
Thrixspermum hainanensis (Rolfe) Schlechter=T. centipeda Loureiro
Tropidia hongkongensis Rolfe (19)
*****Vanda coerulea** Griffith ex Lindley (60)
*****Vanda teres** Lindley (60)
*****Vanda tricolor** Hooker (60)
Vrydagzynea nuda Blume (18)
Zeuxine gracilis (Breda) Blume (16)
Zeuxine leucochila Schlechter (16)
Zeuxine membranacea Lindley (16)
Zeuxine strateumatica (L.) Schlechter (16)

Phytogeographic Significance

The native genera and species of Orchidaceae in Hong Kong had undergone
endless struggles and adjustments to the climatic, edaphic, and biological factors
of the environment before they became established in the area. They are spontaneous
elements of the vegetation. In their genetic constitution, they have inherent factors
fitting for the environmental conditions of this area. After the establishment of their
ancestors in Hong Kong during the Tertiary or earlier, they have inhabited the land
with some adaptive changes. They will continue to live here with necessary modifications or become extinct if certain environmental changes become too destructive or
severe for them to tolerate. As they are now, they serve as good indicators of the
floristic features of the area and give reliable evidences for the interpretation of
phytogeographical relationships.

1. *Indication of a Rich Flora and a Favorable Area*:   The morphological diversity of Orchidaceae in Hong Kong is a prima facie indication of its rich floristic composition. The rich floristic endowment in the area is not limited to Orchidaceae or the large families such as Leguminosae, Gramineae, Compositae, Rosaceae, and Euphorbiaceae. The flora of Hong Kong is rich in the woody families such as Magnoliaceae, Illiciaceae, Hamamelidaceae, Lauraceae, Fagaceae, Moraceae, Theaceae, and Symplocaceae, and the woody climbing families such as the Schizandraceae, Lardizabalaceae, Menispermaceae, Apocynaceae, Asclepiadaceae, and Convolvulaceae. In Hong Kong, a complex assemblage of 1 800 species are known for an area of 404 square miles. This is a much higher area/species ratio than the average calculated by DeWolf for areas of similar size and latitude.[4] Geographically Hong Kong is situated south of the Nan-ling Range (南嶺), on a land mass which has not been subjected to major tectonic and climatic revolutions since the beginning of modern floras. It is part of a land recognized by biogeographers as "favorable area" in the evolution of modern biota.

2. *Reference for Floristic Relationships*:   As mentioned earlier, 98% of the genera of Hong Kong Orchidaceae occur in China. The "world features" column of Table I indicates that Hong Kong shares 66% of its orchid genera with Southeast Asia, 26% with Papuasia, 15% with Africa, 6% with temperate North America and Europe, and 4% with the New World Tropics. There is an obvious gradual reduction of the number of common genera from China to Southeast Asia mainland, thence eastward to Papuasia, Australia and other Pacific islands, and westward to Africa and Madagascar. Hong Kong's geographical position, being less than 100 miles south of the Tropic of Cancer, contributes to the low percentage of genera shared with temperate North America and Europe. Representatives of pantropical genera of orchids are few. This condition accounts for the small number of genera shared with tropical America.

The above floristic relationships as manifested by the Orchidaceae of Hong Kong are indicative of the antiquity of the genera involved, for it takes time for any genus to establish a significant range. The establishment of the ranges showing floristic affinities such as those mentioned above might have taken place through natural means of slow overland dispersal when proper environmental conditions were available. It might have taken place by crossing the barriers in long distance jumps. It is evident that some of the generic ranges of the Orchidaceae of Hong Kong were established through long distance dispersal. *Disperis* and *Cryptostylis* are good examples. *Disperis* is primarily an African genus with 90% of the species occurring in tropical Africa and Madagascar, and one species in each of the following places: Sri Lanka, northern Thailand, Lantau Island of the New Territories of Hong Kong, Taiwan, Luzon of the Philippines, Caroline Islands, New Guinea, and Queensland in Australia. *Cryptostylis* is known from Khasia Hills of northeastern India, Thailand, Hong Kong, Taiwan, thence southward to Malaysia and Sri Lanka, and eastward to Australia and Fiji. Perhaps the long distance transportation of orchid seeds are rendered more effective through attachment to fragments of plant debris carried by

---

[4] DeWolf, G. P. Jr. (1964) On the sizes of floras. Taxon **13**:149–54. Hu, S. Y. (1974) Shea-shore plants of Hong Kong. Journ. Chin. Univ. Hong Kong **2**:331.

strong air current in tropical storms. The partially decayed organic matter serves as a conveyer of the minute seeds and the associated fungi without which the tiny dustlike seeds of orchids do not germinate in nature. The decayed material may also serve as a soft bed ready to absorb moisture in the newly settled habitat.

3. *Endemism at the Species Level*: According to the records in An Alphabetic Enumeration of the Genera and Species[5] of Orchidaceae in China and some recent publication on the orchids of Hong Kong, 32 of the 110 native species in Hong Kong are not known elsewhere. This rate of endemism (30%) at the species levels is surprisingly high. The number may be reduced when material from the belt of "favorable area," which extends from Yunnan eastward to Taiwan, are studied critically in comparison with specimens from Hong Kong. However, high degree of endemism at the species level is not limited to Orchidaceae in Hong Kong. It is characteristic of the flora in the area. This feature is very prominent in the woody genera such as *Illicium*, *Ilex*, and *Camellia*, which have been worked out by modern taxonomists. In the genus *Ilex*, the ratio between the endemic and the wide-spread species for Hong Kong is 1 : 3 (33%), and that in *Camellia* is 1 : 2 (50%).

Evidence accumulated from microscopic observations made on dissection of hundreds of fresh flowers of various orchids and from studying the habitats of different species are conclusive that unique floral structures are developed as isolating mechanisms among species in hitherto little known orchids of the area such as *Cheirostylis*, *Gastrochilus*, and *Zeuxine*. The populations of some species, such as those of *Cheirostylis chinensis* and *C. clibborndyeri*, are widespread, while those of others, such as *C. jamesleungii* and *C. monteiroi*, are restricted. However, the phenomenon of sympatric or contiguous populations of different species of these genera has not been observed. Evidently morphological specialization and geographical isolation are important contributing factors to the high degree of endemism at the species level in Orchidaceae as well as in other families of the flora in Hong Kong.

The native orchids in Hong Kong display phenomenal stability in vegetative characters at the generic or sectional level and spectacular specializations in the floral structures at the species level. In some cases the isolating mechanism between species is chronological as well as morphological and geographical. For example, in the superficial appearance of herbarium specimens of *Cymbidium lancifolium* and *C. maclehoseae*, they look almost alike, for both have lanceolate leaves and lateral inflorescences on ellipsoid pseudobulbs. However, the proximate petals and the flowering habit of *C. lancifolium* in the hot humid summer, contrasted with the widely separated petals and the habit of blooming in the cool dry winter of *C. maclehoseae* seem to indicate an adaptation for different pollinators. Such adaptive features add to the degree of endemism at the species level in orchids.

Conclusion

A systematic summary of the genera of Orchidaceae of Hong Kong reveals that in a territory of 404 square miles, there is an assemblage of species which represent all the subfamilies, tribes, series, 59% of the subtribes and 33% of the genera of Orchidaceae in the entire China. The orchids of Hong Kong display a

---

[5] Hu, S. Y. (1972) The Orchidaceae of China III. Quart. Journ. Taiwan Mus. **24**:41–67.

tremendous range of morphological diversity at the super-generic level. At the infra-generic level, they exhibit a high degree of endemism, for approximately 30% of the species are not known elsewhere. Small number of species and poor population density are two prominent features of the Orchidaceae of Hong Kong. Approximately 60% of the genera have one species each, 12% 3 species each, and 10% 2 species each. Native orchids of Hong Kong are inconspicuous in the landscape because of the predominantly small size of the flowers and the lack of showy color. Their preference for undisturbed habitats and man's ruthless collection and uprooting are also contributing factors for their inconspicuity and population pausity.

The morphological diversity of the Orchidaceae in Hong Kong is a prima facie indication of the rich floristic composition of the area. At the generic level, Hong Kong shares 98% of the orchid genera with China, 66% with Southeast Asia, 26% with Papuasia, 15% with Africa, 60% with temperate North America and Europe, and 4% with tropical America. The genera of Orchidaceae are good indicators of floristic relationships.

In a comment on the discovery of *Gastrochilus holttumianus*, a new species in a hitherto unrecorded genus for Hong Kong, the foremost orchidologist for south-eastern Asia, Prof. R. E. Holttum, Former Director of the Botanic Gardens at Singapore wrote, "I note that it was found on Victoria Peak, Hong Kong Island. It is strange that no one found it previously."[6] Sitting in a cross harbor ferry and looking up at the massive northern flank of Victoria Peak with its perpendicular igneous rocks exposed from the adjacent woods, one cannot help from wondering whether the steep slope has ever been botanized, and how many orchids there are, waiting on the rock and in the forest for discovery.

---

[6] Hu, S. Y. & Gloria Barretto. (1975) New species and varieties of Orchidaceae from Hong Kong. Chung Chi Journ. **14**:34.

IV
Origin and Meaning of the Generic Names of Hong Kong Orchids

In theory, the name of a genus may be derived from any source. In practice, botanists draw heavily upon Greek or Latin for the composition of the generic name. Fifty of the 64 generic names of Hong Kong orchids are of Greek origin, 6 are from Latin, and one is from Sanskrit. The remaining 7 genera are named after persons.

The Chinese equivalents of these generic names are taken from several sources. These are: (1) 5 from the Chinese classics; (2) 14 from the existing vernacular names; (3) 29 from a translation of names of Greek or Latin origin; and (4) 16 from recent creations of Chinese orchidologists.

In this chapter, the origin and meaning of the generic names of Hong Kong orchids and their Chinese equivalents are explained.

Acampe Lindley, Fol. Orch. Acampe 5. 1853; Greek *akampes* = rigid, possibly referring to the stout rigid root, stem and leaves of the type species.

Ts'ui-hua 脆花 "Fragile Orchid," a classical name, referring to the delicate flowers of *A. multiflora*.

Acanthephippium Blume, Bijdr. 353. 1825; Greek *akantha* = thorn, and *ephippion* = saddle, referring to the saddle-like lateral lobes of the apical portion of the lip.

T'ang-hua Lan 罈花蘭 "Jug-flowered Orchid," a descriptive term alluding to the shape of the flower, which form a tube swollen on one side, giving it the appearance of a jug (Fig. 50).

Ania Lindley (Ascotaenia Ridley), Gen. Sp. Orchid. 129. 1831; Greek *ania* = trouble, indicating the problem of the taxonomic relationship of this species.

An Lan 安蘭 "Peace Orchid," a translation into Chinese of the sound of Lindley's generic name, *Ania*.

Anoectochilus Blume, Bijdr. 411. 1825; et Tab. 15. 1825 (*Anectochilas*); Greek *anoektos* = open, and *cheilos* = lip, referring to the infolding margins of the channeled claw which leads to the conic spur of the bending lip.

K'ai-ch'un Lan 開唇蘭 "Open Lip Orchid," a translation of the meaning of the generic Greek name.

Aphyllorchis Blume, Tab. Orch. t. 77. 1826; Greek *aphyllos* = leafless, and *orchis* = orchid, referring to the leafless saprophytic condition of the species.

Wu-yeh Lan 無葉蘭 "Leafless Orchid," a translation of the meaning of the Greek generic name.

Appendicula Blume, Bijdr. 297. 1825; et Tab. 40. 1825; Latin *appendicula* = little appendage, referring to the hanging appendage in the sac of the lip.

Niu-tzu Lan 牛齒蘭 "Ox-teeth Orchid," a descriptive term used by Tang and Wang to call attention to closely arranged leaves which has a fanciful resemblance of the teeth of ox.

Arachnis Blume, Bijdr. 365. 1825; Greek *arachne* = spider, referring to the long and narrow sepals and petals which emphasizes the flowers resemblance to a spider.

Chih-chu Lan 蜘蛛蘭 "Spider Orchid," a translation of the meaning of the Greek generic name.

Arundina Blume, Bijdr. 401. 1825; Latin *arundo* = reed, referring to the reed-like habit.

Chu-yeh Lan 竹葉蘭 "Bamboo Leaf Orchid," a Chinese vernacular name which describes the bamboo-like leaves of a species in South China (*A. chinensis*, Fig. 29).

Bletilla Reichenbach fil. in Fl. Serr. Jard. **8**: 246. 1853; a diminutive of *Bletia*, a genus of orchids in the New World.

Po-chi 白及 "White Chicken," a classical term used in Chinese Materia Medica. The name carries the sound of a vernacular name of *Bletilla striata* used in Szechwan and Kweichow. The cultivated form of this species in western and central China produces a fleshy white tuber which resembles a small white chicken (白雞兒 = Po-chi êrh), and this is the name used locally. In prescribing the medicine, the practitioners have used the characters 白及 (Po-chi), which give the sound but miss the original meaning.

Brachycorythis Lindley, Gen. Sp. Orch. 363. 1828; from Greek *brachys* = short, and *korys* = helmet, referring to the hook formed by the short dorsal sepal connivent with the small petals.

Tuan-k'uei Lan 短盔蘭 "Short Helmet Orchid," a translation of the meaning of the generic name; appeared first in Quart. Jour. Taiwan Mus. **25**:59. 1972.

Bulbophyllum Thouars, Orch. Afr. t. 93–97, 99–110, 1822; Greek *bolbos* = bulb, and *phyllon* = leaf, referring to the fleshy leaf terminal to a pseudobulb.

Shih-tou Lan 石豆蘭 "Rock-bean Orchid," a vernacular name describing both the epiphytic habit and the shape of the pseudobulbs which resemble some large beans.

Calanthe R. Brown in Bot. Reg. **7**: t. 573. 1821; Greek *kalos* = beautiful, and *anthe* = bloom, alluding to the beautiful flower of the species.

Hsia-chi Lan 蝦脊蘭 "Shrimp's Back Orchid," a name employed by Tang and Wang to refer to the slightly curved column of some species.

Cattleya Lindley, Coll. Bot. tt. 33, 37. 1821; named for William Cattley, a business-man of London and a collector of rare plants in the early part of the nineteenth century.

Pu-tai Lan 布袋蘭 "Calico Sac Orchid," a common Chinese name for the flowers of cultivars cultivated in Taiwan and Hong Kong, referring to the lip with ruffled margin.

Cephalantheropsis Guillaumin in Bull. Mus. Hist. Nat. Paris II. **32**:188. 1960; a combination of *Cephalanthera* = the generic name of orchids, and Greek *opsis* = resembling, referring to the resemblance of the flowers to those of *Cephalanthera*.

Hsiao T'ou-jui Lan 肖頭蕊蘭 "Imitating Cephalanthera," a translation of the meaning of the generic name.

Cheirostylis Blume, Bijdr. 413. 1825; et Tab. 16. 1825; Greek *cheir* = hand, and *stylis* = style, referring to the hand-shaped clinandrium.

Ch'a-chü Lan 叉柱蘭 "Forked Pillar Orchid," a descriptive term used by Tang and Wang, alluding to the arms of the column.

Cirrhopetalum Lindley in Bot. Reg. **10**: t. 832. 1824; Greek *kirrhos* = orange-tawny, and Latin *cirrus* = fringe, referring to the golden fringe of the petals of some species.

Chuan-pan Lan 捲瓣蘭 "Curled Sepal Orchid," a descriptive term used by Tang and Wang which refers to the twisted sepals.

Cleisostoma Blume, Bijdr. 362. 1825; Greek *kleistos* = closed, and *stoma* = mouth, referring to the callus on the back wall of the mouth of the spur.

Fêng-k'ou Lan 封口蘭 "Closed Mouth Orchid," a translation of the generic name of Greek origin.

Coelogyne Lindley, Collect. t. 33. 1821; Greek *koilos* = hallow, and *gyne* = pistil, referring to the stigma deeply excavated on the column.

Pi-mu Lan 貝母蘭 "Mother-of-pearl Orchid," a vernacular name describing the shape of the pseudobulbs, and their medicinal value.

Cryptostylis R. Brown, Prodr. 317. 1810; Greek *kryptos* = hidden, and *stylis* = style, referring to the hidden position of the column in a depression at the basal portion of the lip.

Yên-chu Lan 隱柱蘭 "Hidden Column Orchid," a translation of the technical generic name of Greek origin.

Cymbidium Swartz in Nov. Act. Soc. Sci. Upsal. **6**:70. 1799; Greek *kymbes* = boat-shaped referring to the basal portion of the lip (Fig. 55).

Lan 蘭 "Orchid," the basic name for the orchid family.

Dendrobium Swartz in Nov. Act. Soc. Sci. Upsal. **6**:82. 1799; Greek *dendron* = tree, and *bios* = life, referring to the epiphytic habit of some species of the genus.

Shih-hu 石斛 "Rock Living," a descriptive term adopted from the vernacular name that refers to the habitat of some species (Fig. 28). The stems of many species are used in medicine. City practitioners have erroneously applied a homonymic character 斛, in place of 活, and made the name meaningless. *Shih-hu* is one of the basic terms and it does not carry the character "Lan." The Chinese generic name for *Desmotrichum*, *Chin-shih-hu* 金石斛, is built on this basinym.

Diploprora Hooker f., Fl. Brit. Ind. **6**:26. 1890; Greek *diplous* = double, and *prora* = prow, referring to the tentaculate apical lobes of the lip.

Shê-shê Lan 蛇舌蘭 "Snake Tongue Orchid," a vernacular name describing the tip of the lip which resembles the tongue of a snake.

Disperis Swartz in Vet. Acad. Handl. Stockh. 21. 218. 1800; Greek *dis* = twice, and *pera* = wallet or sac, referring to the pouches formed at the middle of the bending sepals.

Shuang-tai Lan 雙袋蘭 "Double Sac Orchid," a translation of the generic name of Greek origin, first appeared in Quart. Jour. Taiwan Mus. **26**:376. 1973.

Epidendrum Linnaeus, Syst. Nat. ed. 10. **2**:1246. 1759, p. p., non Linn. 1753, *nom. cons.*, Greek *epi* = on, and *dendron* = tree, alluding to the epiphytic habit of the species.

Shu-shêng Lan 樹生蘭 "Tree Growing Orchid," a translation of the generic name of Greek origin.

Eria Lindley in Bot. Reg. **11**: t. 904. 1825; Greek *erion* = wool, in allusion to the woolly flowers.

Mao Lan 毛蘭 "Hairy Orchid," a descriptive term used by Tang and Wang, referring to the hairy scape and flower. Some authors prefer to call it *Yung Lan* 絨蘭 "Velvet Orchid," which refers to the same peculiar character of some species of the genus.

Eulophia R. Brown in Bot. Reg. **8**: t. 686. 1823; Greek *eu* = well, and *lophos* = plume, referring to the crest on the lip.

Mei-kuan Lan 美冠蘭 "Beautiful Crown Orchid," a translation of the technical term of Greek origin.

Gastrochilus D. Don, Prodr. Fl. Nepal. 32. 1825; Greek *gaster* = belly, and *cheilos* = lip, referring to the belly-shaped lip (Fig. 67).

Tu-ch'un Lan 肚唇蘭 "Belly Orchid," a translation of the generic name of Greek origin.

Geodorum Jackson in Andrews, Bot. Repos. **10**: t. 626. 1810; Greek *ge* = earth, and *doron* = gift, referring to the terrestrial habit.

Ti-pao Lan 地寶蘭 "Treasure of Earth Orchid," a translation of the technical name of Greek origin: "gift of the earth."

Goodyera R. Brown in Aiton, Hort. Kew. ed. 2. **5**:197. 1813; named for an English botanist John Goodyer (1592–1664).

Pan-yeh Lan 斑葉蘭 "Variegated Leaf Orchid," a vernacular name adopted for the genus by Tang and Wang, referring to the beautiful patterns of the leaves of some species.

Habenaria Willdenow, Sp. Pl. **4**:44. 1805; Latin *habena* = reins, referring to the strap-like median longitudinal band of the petals and lip of some species.

Yü-fêng Hua 玉鳳花 "Jade-phoenix Flower," a descriptive name used in the Chinese classics for the white flower with an elongated spur that resembles the long tail of a bird.

Hetaeria Blume, Fl. Jav. 84. 1858; Greek *hetaireia* = companionship, referring to its close relationship with *Goodyera* and allied genera.

Pan Lan 伴蘭 "Companion Orchid," a translation of the generic name of Greek origin.

Liparis C. L. Richard in Mém. Mus. Hist. Nat. Paris **4**:52. t. 5. fig. 10. 1818; Greek *liparos* = greasy, referring to the shiny surface of the leaves of some species.

Yang-erh Lan 羊耳蘭 "Goat-ear Orchid," adopted from a vernacular name which refers to the shape of the leaves of some species.

Ludisia A. Richard in Dict. Class. Hist. Nat. **7**:437. 1825; Richard did not record the meaning when he proposed this generic name, perhaps for one of his friends.

Hsüeh-yeh Lan 血葉蘭 "Blood Leaf Orchid," a vernacular name referring to the dark red color of the leaves of *L. discolor*, a species in South China.

Lycaste Lindley in Bot. Reg. **29**(Misc.):14. 1843; Greek *Lycaste* = a goddess of the mountains, forests, meadows, or waters in Greek mythology.

Li-hsien Lan 麗仙蘭 "Beautiful Fairy Orchid," a translation partially of the sound and partly of the meaning of the generic name of Greek origin, first appeared in Quart. Jour. Taiwan Mus. **27**:432. 1974.

Malaxis Solander ex Swarz, Prodr. Veg. Ind. Occ. 119. 1788; Greek *malaxis* = softening, referring to the texture of the leaves.

Chao Lan 沼蘭 "Marsh Orchid," adopted from a classical term describing the habitat of some species.

Manniella Reichenbach fil., Otia Bot. Hamb. **2**:109. 1881; named for Gustaf Mann, explorer of the mountains of Cameroon.

Man-shih Lan 滿氏蘭 "Mann's Orchid," a translation of the Latinized generic name, a new introduction here, the genus being reported from Hong Kong recently.

Microcoelia Lindley, Gen. Sp. Orch. 60. 1830; Greek *mikros* = small, and *koilia* = abdomen, referring to the very small flower with short and stout column, the entire front of which is occupied by a concaved stigma.

Hsi Kên Lan 細根蘭 "Delicate Root Orchid," describing the small aphyllous plant with numerous cord-like roots.

Miltonia Lindley in Bot. Reg. **23**: t. 1967. 1837; named for Viscount Milton (1786–1857), a British orchidophile of the nineteenth century.

Mai-shih Lan 麥氏蘭 "Milton's Orchid," a translation of the Latinized generic name, first used in Quart. Jour. Taiwan Mus. **27**:437. 1974.

Mischobulbum Schlechter in Repert. Sp. Nov. Fedde Beih, **1**:98. 1911; Greek *mischos* = stalk, and *bolbos* = bulb, referring to the petiole-like pseudobulb.

Ch'iu-ping Lan 球柄蘭 "Bulb Stalk Orchid," a translation of the technical term of Greek origin.

Nephelaphyllum Blume, Bijdr. 372. 1825; et Tab. 22. 1825; Greek *nephela* = cloud, and *phyllon* = leaf, referring to the brownish, greenish, mottled leaves (Fig. 31).

Yün-yeh Lan 雲葉蘭 "Cloud-leaf Orchid," a translation of the meaning of the technical term of Greek origin.

Nervilia Commerson ex Gaudichaud in Freyc. Voy. Bot. 422. t. 35. 1826; Latin *nervus* = vein, referring to the prominent veins of the leaves.

Yü Lan 芋蘭 "Taro Orchid," a classical term adopted for this genus by Tang and Wang, referring to the corms of *N. fordii* (Fig. 12b–e).

Odontoglossum Humboldt, Bonpland & Kunth, Nov. Gen. Sp. **1**:350. t. 85. 1815; Greek *odous, odonto* = tooth, and *glossa* = tongue, referring to the tooth-like projections on the disk of the lip.

Ya-shê Lan 牙舌蘭 "Toothed Tongue Orchid," a translation of the generic name of Greek origin.

Oncidium Swartz in Vet. Acad. Handl. Stockh. **21**:239. 1800; Greek *onkos* diminutive = tumor, referring to the warty callus of the lip.

Liu-ch'un Lan 瘤唇蘭 "Warty Lip Orchid," a translation of the generic name of Greek origin.

Ornithochilus Wallich ex Lindley, Gen. Sp. Orch. 242. 1833; Greek *ornithos* = bird, and *cheilos* = lip, referring to the divaricate lobes of the lip which resemble a flying bird.

Yü-ch'un Lan 羽唇蘭 "Feather Lip Orchid," a term used by Tang and Wang to describe the fringed lip which resembles a bird's feather.

Pachystoma Blume, Bijdr. 376. 1825; Greek *pachys* = thick, and *stoma* = mouth, referring to the fleshy lip covered by papillose callosities.

Fêng-k'o Lan 粉口蘭 "Powder Mouth Orchid," a descriptive term appearing repeatedly in recent publications, probably referring to the papillose disk of the lip of *Pachystoma chinense*.

Paphiopedilum Pfitzer, Morph. St. Orch. 11. 1886, *nom conserv.*, Greek *Paphia* = Paphos, epithet of Venus, and *pedilon* = sandal, referring to the slipper-shaped lip.

Tou Lan 兜蘭 "Sac Orchid," a name referring to the saccate lip of the flowers in this genus. Some authors prefer to use the term *Tou-shê Lan* 兜舌蘭 "Saccate-lip Orchid."

Pecteilis Rafinesque, Fl. Tellur. **2**:37. 1836; Greek *pectein* = to comb, referring to the pectinate lateral lobes of the lip.

Po-tieh Hua 白蝶花 "White Butterfly Flower," adopted from a vernacular name describing the appearance of the flower which resembles a white butterfly.

Peristylus Blume, Bijdr. 404. 1825; et Tab. 30. 1825; Greek *peri* = around, and *stylos* = column, referring to the glands on the sides of the column (Fig. 11ꜰ).

K'uo-jui Lan 闊蕊蘭 "Broad Stamen Orchid," a term proposed by Tang and Wang to describe the broad rudimentary stamens.

Phaius Loureiro, Fl. Cochinch. 529. 1790; Greek *phios* = gray, referring to the gray hue on the outside of the sepals and petals.

Ho-ting Lan 鶴頂蘭 "Stork's Crown Orchid," adopted from the vernacular name of *Phaius tankervilliae*, a favorite of gardeners in South China. The name refers to the shape of the white flower-bud. At a certain stage of its development, the bud resembles the head of a stork (Fig. 51).

Phalaenopsis Blume, Bijdr. 274. 1825; Greek *phalaina* = moth, and *opsis* = appearance, referring to the shape of the flower which resembles a butterfly or a moth.

Tieh Lan 蝶蘭 "Butterfly Orchid," a translation from the technical term of Greek origin.

Pholidota Lindley ex Hooker Exot. Fl. **2**: t. 138. 1825; Greek *pholidotos* = scaly, referring to the imbricate scales of the young inflorescence.

Shan T'ao Lan 山桃蘭 "Mountain-peach Orchid," a term adopted from a vernacular name of certain species, describing the pseudobulb which resembles the shape of a peach. The word *Shan* generally stands for mountain or hill. In the colloquial expressions of southern and western China it may mean "wild form," especially when it is used in connection with plants. Some authors prefer to use the term *Shih Hsien-t'ao* 石仙桃 (Fairy's Peach-on-the-rock).

Platanthera L. C. Richard in Mém. Mus. Hist. Nat. Paris **4**:48. 1818; Greek *platys* = broad, and *anthera* = anther, referring to the short and wide anther.

Chang-chü Lan 長距蘭 "Long Spur Orchid," a descriptive term used by Tang and Wang in allusion to the elongated spurs. Some authors prefer to call this genus *Fêng-tieh Lan* 粉蝶蘭 "Cabbage-butterfly Orchid."

Renanthera Loureiro, Fl. Cochinch. 521. 1970; Latin *renes* = kidney, and Greek *anthera* = anther, referring to the reniform anther of *R. coccinea*, a favorite of Chinese orchid collectors.

Huo-yên Lan 火焰蘭 "Tongue of Flame Orchid," a term adopted from the vernacular name which refers to the large panicle of conspicuous red flowers, fancifully alluding to the tongue of flame.

Robiquetia Guadichaud in Freycinet, Voy. Bot. 426. t. 34. 1826; named for a French phytochemist M. Pierre Robiquet, the discoverer of caffine and morphine.

Lu-pi Lan 露壁蘭 "Exposed Precipice Orchid," a translation of the first two syllables of the generic name. The Chinese characters convey the habitat of *R. succisa*, a species growing on exposed precipices of Hong Kong and the warmer part of China.

Sophronitis Lindley in Bot. Reg. **14**: t. 1147. 1828; Greek *sophron* = modest, diminutive of the genus *Sophronia*, referring to the close relationship to the modest orchid of inconspicuous habit.

Chêng Lan 貞蘭 "Modest Orchid," a translation of the meaning of the generic name of Greek origin.

Spathoglottis Blume, Bijdr. 400. 1825; Greek *spathe* = spathe, and *glotta* = tongue, referring to the broad midlobe of the lip.

Pao-shê Lan 苞舌蘭 "Bract-tongue Orchid," a translation of the technical term of Greek origin.

Spiranthes L. C. Richard in Mém. Mus. Hist. Nat. Paris **4**:50. 1818; Greek *speira* = coil, and *anthos* = flower, referring to the spirally arranged flowers of the spike.

Shou Ts'ao 綬草 "Tassel Grass," a name adopted from a classical term referring to the grass-like appearance and the crisped, tassel-like lip of S. *sinensis* Ames, a widespread weed in depressions on lawns.

Tainia Blume, Bijdr. 354. 1825; Greek *tainia* = fillet, referring to the elliptic leaf on an elongated petiole.

Têng Lan 鄧蘭 "Têng Orchid," a translation into a Chinese character with sound close to the first syllable of *Tainia*.

Thelasis Blume, Bijdr. 385. 1825; et Tab. 75. 1825; Greek *thele* = nipple, possibly referring to the small flower bud before it opens.

Ai-chü Lan 矮柱蘭 "Short Column Orchid," a term used by Tang and Wang for the very short column almost completely covered by the stigma.

Thrixspermum Loureiro, Fl. Cochinch. 519. 1790; Greek *thrix* = hair, *sperma* = seed, referring to the minute winged seeds which appear like hairs within the first opened capsule.

Po-tien Lan 白點蘭 "White-spotted Orchid," a term adopted from a vernacular name for the white lip with a dark central spot which represents the stigma in the flower of the type species, *T. centipeda*.

Tropidia Lindley in Bot. Reg. **19**: t. 1618. 1833; Greek *tropideion* = keel, referring to the boat-shaped lip.

Chü-ching Lan 竹莖蘭 "Bamboo-stick Orchid," a descriptive term employed by Tang and Wang because of the resemblance of the stem to the culm of a bamboo.

Vanda Jones in Asiatic Researches **4**:302. 1795; Sanskrit *Vanda*, a vernacular name for *Vanda Roxburghii*, a species first known from Bengal. Jones reported that "vanda" is the local name of air-plants which live without soil or water. People in Bengal call *Loranthus* "baculavanda," and *Viscum* "vanda."

Wan-tai Lan 萬帶蘭 "Ten Thousand Ribbon Orchid," a term very cleverly proposed by Tang and Wang. It may be interpreted as a translation of the sound of the generic name of Sanskrit origin, or it may be taken also as a descriptive term referring to the numerous ribbon-like leaves of many species in the genus.

Vrydagzynea Blume, Orch. Archip. Ind. 71. t. 17–22. 1857; named for a Dutch pharmacologist, T. D. Vrydag Zynen.

Er-wei Lan 二尾蘭 "Two Tailed Orchid," a descriptive term used by Tang and Wang, referring to the tail-like stipitated glands pointing backwards to the spur of *V. nuda* (Fig. 12ᴇ).

Zeuxine Lindley, Orch. Scel. 9. 1826; Greek *zeuxis* = yoking, referring to the partial union of the lip and column.

Hsien-chu Lan 線柱蘭 "Thread-column Orchid," a descriptive term used by Tang and Wang, calling attention to the short column with elongated arms.

Glossary

Acaulescent (無莖). Plant without evident stem and with a rosette of radical leaves on a very short erect rhizome, as in *Manniella* and *Cryptostylis*.

Acuminate (漸尖). Tapering to a slender point.

Acute (急尖). Sharply pointed, but not drawn out to a slender point.

Adherent (附着). Coming into close and face-to-face contact, such as the odd sepal and petals of *Goodyera* and *Disperis*.

Agglutinate (膠合的). United by a glue as the pollen grains of most orchids.

Androecium (雄蕊). The stamens; the male floral part of a flower.

Anthesis (花期). The period in which the flowers open.

Aphyllous (無葉). Without leaves, such as *Microcoelia* and *Aphyllorchis*.

Apiculate (具細尖). A short, sharp, but not stiff point at the tip of an organ.

Appendage (附屬物). An attached secondary part, such as the flap at the base of the lip of *Appendicula*.

Arcuate (弧曲). Curved or bending forward, such as the column of *Liparis*.

Articulate (具關節). Jointed; having a node or point where separation may take place naturally, such as the leaves of *Acampe* and *Dendrobium*.

Auricle (耳器). Ear-like appendage on the column, glandular in *Habenaria*, petaloid in *Oncidium*, or at the base of the lip as in *Malaxis*.

Auriculate (耳狀). Bearing ear-like processes.

Band-shaped (帶狀). Flat and elongated leaves parallel along the margin, such as that of *Acampe*.

Bi (二). A Latin prefix signifying two or twice, as bilobed (二裂), bicornute (具二角), or bicuspidate (具二硬尖).

Bifurcate (二义的). Forked, or two-pronged.

Blade (葉片). The expanded part of a leaf, = lamina.

Bract (苞片). A modified leaf, especially the one from the axil of which a flower is developed.

Calcarate (有距的). Spurred.

Callus (胼胝). A knob-like excrescence on the disc, in the spur, or occasionally outside the midlobe of the lip, such as in *Liparis, Cleisostoma,* and *Arachnis*.

Canaliculate (具溝的). Channelled, as the space between the keels on the disc of *Cymbidium*.

Capsule (蒴果). A dry dehiscent fruit developed from a multicarpellate ovary, as in all orchids.

Cartilaginous (脆骨質). Hard and tough, leathery, as the inner wall of orchid fruits.

Cataphyll (芽苞葉). The scale leaves of rhizomes or the lower sheaths of an erect stem.

Caudicle (花粉塊柄). The slender stalk-like appendage of the pollinia.

Chin (跟). A term formerly used in place of mentum, seldom used now.

149

Cilia (sing., cilium 睫毛). Long hairs along the margin, as in the petals of some species of *Cirrhopetalum*.

Claw (爪). The narrow portion of a lip with expanded apex, such as that of *Zeuxine* or *Cheirostylis*.

Cleistogamous (閉花受精的). Having close fertilization within unopened flowers.

Clinandrium (藥座). A cavity in the apex of the column containing the anther in orchids

Column (蕊柱). The organ that bears the anther and stigma in orchids.

Concave (凹的). Curved along the margin and producing a hollow space in between as the lip of *Geodorum*.

Connivent (靠合). Coming into contact, such as the lateral sepals of *Cirrhopetalum*, or the odd sepal and petals of *Goodyera*.

Coriaceous (革質). Leathery, referring to the thick leaves of many species.

Corm (球莖). A solid sub-spherical underground stem, covered by scales or with rings indicating degenerated scales.

Cornute (角形). Having the shape of horns, as the anther of *Tainia, Nephelaphyllum*, and *Mischobulbum*.

Cristate (鳥冠狀). Like the crest of a bird, as the disc of the lip of *Oncidium*.

Cucullate (盔狀). Helmet-shaped, as the anther of *Acanthephippium*.

Cultivar (小品). Clones selected from hybridization and multiplied by vegetative means, as *Miltonia* ''Belt Field.''

Cuneate (楔形). Wedge-shaped.

Cupular (杯狀). Having the form of a shallow-cup, such as the clinandrium of *Zeuxine*.

Cuspidate (具硬尖). Having a cusp, as the apex of the column of *Appendicula* or *Bulbophyllum*.

Cyathiform (盃狀). Cup-shaped, as the clinandrium of *Anoetochilus*.

Cymbiform (舟狀). Boat-shaped, as the bracts of many orchids.

Diandrous (雙葯的). Orchids with two anthers, as in *Paphiopedilum* (Fig. 2A, C).

Disc (花盤). The portion of the lip between the lateral and the middle lobes.

Emarginate (微缺的). Having a notch at the tip.

Endemic (特產). Native or local to a restricted area.

Endemism (特有現象). A condition by which a taxon has a restricted area of distribution.

Entire (全緣). With even margin, not lobed or toothed.

Ephemeral (短壽). Short-lived, existing only a day, as the flower of *Thrixspermum*.

Epiphyte (附生植物). A plant growing on a rock or other plant.

Epiphytic (附生的). Growing upon object such as a rock or a tree.

Equitant (套摺). Folded over as if astride, such as the leaves of *Miltonia*.

Erect (直立). Upright.

Erostrate (無喙). Without a beak, as the fruit of *Cryptostylis*.

Falcate (鐮形). Sickle-like, with curved margin, as the petal of *Cirrhopetalum delitescens*.

Fimbriate (流蘇形). Fringed, as the margin of the lip of *Coelogyne fimbriata*.

Flabelliform (扇狀). Having the form of a fan, as the inflorescence of *Cirrhopetalum*.

Foot (蕊柱足). The forward extention of the base of the column on which the lip is attached, as in *Eria* and *Tainia*.

Forked (分义). Branched into a fork, as the rostellum of *Goodyera*.

Fornicate (拱起的). Arched, as the odd sepal of *Cirrhopetalum*.

Fringed (流蘇狀). Furnished with a fringe, as the lip of *Anoectochilus*.

Fusiform (紡錘狀). Spindle-shaped, as the capsules of *Malaxis*.

Galea (盔瓣). A helmet-shaped structure formed by the odd sepal and petals, such as in *Goodyera*.

Gibbous (淺囊狀). With a pouch-like enlargement, as the base of the lip of *Nervilia*.

Gland (腺). A secreting structure.

Globose (球形). Spherical, globular.

Granular, granulose (顆粒狀). Composed of or appearing as if covered by minute grains.

Gynandrium (蕊柱). The organ that bears the anther and stigma, = column.

Gynandrous (雌雄蕊合體的). With the stamens adnate to the pistil, characteristic of the orchid family.

Gynoecium (雌蕊). A collective term for the pistil or pistils of a flower.

Hood-like (盔狀). A protective cover formed by the coherent odd sepal and the petals, as in *Goodyera* and *Peristylus*.

Incumbent (內曲的). Resting or leaning upon, such as the incumbent anther.

Incurved (內彎). Curved inward.

Inflexed (內捲). Turning inward such as the margin of the lip of *Cheirostylis, Hetaeria,* and *Zeuxine*.

Lamellate (褶片狀). Made up of thin plates, such as the lip of *Bletilla*.

Laminiform (葉片狀)*. Flat, like a lamina, such as the appendage attached to the pollinia of *Bletilla* or *Cattleya*.

Lanceolate (披針形). Lance-shaped; narrow and tapering to the apex.

Ligulate (舌狀). Tongue-like, such as the lip of *Liparis*.

Linear (線形). Narrow and several times longer than wide.

Linguiform (舌狀). Having the form of a tongue.

Lip (唇). A modified petal of an orchid (labellum).

Lunate (新月形). Crescent-shaped, as the calli of some *Zeuxine*.

Maculate (具斑點). Blotched or spotted, as the sepals and petals of *Acampe* and *Oncidium*.

Mentum (跟). The structure formed by the adnation of the bases of the lateral sepals to the column-foot, as in *Eria* and *Bulbophyllum*.

Mobile (能活動). Movable, as the lip of *Bulbophyllum*.

Monandrous (單蕊的). Orchids with one anther as in *Habenaria* and *Phaius* (Figs. 2B, C and 3A–B).

Monochasial (單歧的). Pertaining to a cyme with unpaired flowers.

Monopodial (單軸的). Having unlimit apical growth, as the stem of *Acampe*.

Niche (境所)*. A locality with an ecological condition suitable for the establishment and the perpetuation of the life of a taxon.

Notch (凹). A V-shaped indentation.

Nutant (俯垂). Nodding, drooping, as the raceme of *Geodorum*.

Oblanceolate (倒披針形). When the broadest part of a lanceolate blade is nearer the tip than the base.

Oblong (長圓形). At least twice as broad as long, with the middle portion of the margin more or less parallel.

Operculum (蕊蓋). The cover of the pollinia; anther cap of some authors.

Orbicular (正圓形). A flat body with circular outline.

Orifice (穴孔). The aperture of a spur.

Ovary (子房). The part of a pistil which contains the ovules.

Panduriform, pandurate (提琴形). Fiddle-shaped as the lip of *Spathoglottis*.

Pedicel (花梗). The stalk supporting a single flower.

Peduncle (總梗). A stout flower stalk, supporting a cluster of flowers.

Pentacyclic (五輪列的). The floral parts in five whorls (Fig. 2A, B).

Persistent (宿存). Not falling off at maturity, as the column of *Cymbidium*, or the perianth of *Spiranthes*.

Petals (花瓣). Two segments of the inner whorl of the perianth of orchid.

Plicate (具褶). Folded into plates, as the leaves of *Acanthephippium* and *Malaxis*.

Pollinia (sing., pollinium 花粉塊). The agglutinated mass of pollen grains.

Powdery (粉狀). Pollen grains of fine texture.

Produced (引長的). Growing outward; elongated.

Pseudobulb (假鱗莖). A thickened and bulb-like internode, as in *Eria* and *Bulbophyllum*.

---

* New translations, not known to have been used elsewhere.

Pulvinate (墊狀). Cushion-like.

Pyriform (梨狀). Having the shape of a pear, as the pollinia of *Eria rosea*.

Radical (根生). Said of leaves born in a rosette on short erect rhizome concealed by the leaf-bases.

Reclinate (前後對摺). Bent down or falling back from the apex.

Resupinate (轉向). The phenomenon of twisting of the ovary for 180° in a fully developed flower bud of orchid, consequently placing the lip at the lowermost position, as in *Cymbidium* and *Phaius* (Fig. 2ᴇ).

Retuse (微凹). Notched at the apex as column of *Platanthera*, and leaf of *Arachnis*.

Rhizome (根莖). Creeping or lumpy stem of some orchids as in *Goodyera* and *Cymbidium*.

Rostellum (蕊喙). A thin membranous or fleshy projection of the column between the anther and the stigma or stigmas.

Rostrate (喙狀). Beak-like, referring to capsules with a persistent column.

Rotundate (略圓). Roundish.

Saccate (囊狀). Bag-shaped, as the base of the lip of *Vanda*.

Saprophytic (腐生的). Living on dead organic matter, as species of *Eulophia*.

Saxicolous (石活). Growing on rocks, as *Eria* and *Dendrobium*.

Scape (花葶). A leafless peduncle.

Sectile (方格狀). Pollinia with tessellate appearance.

Septate (具隔膜). Divided by septa, as the thecae of *Cattleya*.

Seta (pl., setae 剛毛). A bristle-like hair, as that on the apex of the petal of *Cirrhopetalum delitescens*.

Sheath (鞘). Tubular portion of the petiole.

Spike (穗狀花序). An elongated cluster of sessile flowers as those of *Spiranthes* and *Peristylus*.

Spur (距). A hollow slender usually nectariferous extension of the lip, such as *Habenaria* and *Peristylus*.

Staminal (雄蕊的). Pertaining to the stamen.

Staminode (退化雄蕊). A sterile stamen, or an organ growing in the place of a stamen, such as the butterfly-like or discoid structure at the end of the column of *Paphiopedelium*.

Stipe (柄). The stalk on which the pollinia and the viscidium are attached, as in *Vanda, Cleisostoma*, and *Cymbidium*.

Striate-sulcate (具脊溝). Having ridges and grooves.

Subterete (近柱形). Almost cylindric.

Subumbelliform (近傘形)*. A raceme with some flowers arranged close together.

Symbiosis (共生). The phenomenon of living together in intimate association and of being mutually beneficial.

Symbiotic (共生的). Living beneficially together.

Sympatric (同鄉). Taxa sharing the same general geographical range.

Sympodial (合軸的). Stems with limited apical growth, as *Coelogyne* and *Cymbidium*.

Synsepalum (合萼). The combined structure of the lateral sepals of *Paphiopedilum*, normally hidden behind the lip.

Terete (圓柱形). Cylindric.

Terrestrial (陸生的). Used for plants growing on the ground.

Tessellate (棋盤狀). In form of squares, as the dark and pale colors on some leaves of the slipper orchids; also of the pollinia of *Goodyera*.

Theca (pl., thecae 葯室). The pollen chamber of an anther.

Tricarpellate (三心皮的). An ovary consisting of three carpels, characteristic of the orchid family.

Trilobed (三瓣的). Having a lip with two lateral lobes and a midlobe, as in *Habenaria*.

Trimerous (三基數的). The floral parts with three segments in each whorl, a basic pattern in the structure of the flowers of orchids (Fig. 2ᴀ, ʙ, ꜰ).

Truncate (截形). As though cut off at the end.

Tuberculate (瘤狀). Warty.

Unguiculate (爪狀). Clawed; having a slender narrow base, as the midlobe of the lip of *Spathoglottis* and *Phalaenopsis*.

Urceolate (罎狀). Tubular and contracted at the mouth as the sepals of *Acanthephippium*.

Velamen (pl., velamina 根被). A protective spongy multicellular epidermal layer of dead cells on the outside of the root of orchid, which gives it a silvery-gray color.

Venose (顯脈). Having conspicuous veins as the plicate-venose leaves of *Phaius, Calanthe,* or *Geodorum.*

Ventricose (一面臌). Swelling unequally, or inflated on one side, as the lip of *Goodyera.*

Viscidium (黏盤). Viscid disc of orchid.

Waxy (蠟質的). Shiny and smooth, as the pollinia of *Cymbidium* and *Vanda.*

Bibliography

Ames, O. Orchidaceae. Vol. I. 1–156. 1905; II. 1–288. 1908; III. 1–99. 1908; IV. 1–287. 1910; V. 1–271 (Genera & Species of Philippine Orchids). 1915; VI. 1–335 (Orchids of Mount Kinabala). 1920; VII. 1–174. 1922.

Anonymous. Check List of Hong Kong Plants. x + 115. Government Press, Hong Kong. 1974.

Bateman, J. The Orchidaceae of Mexico and Guatemala. 12 + plates 40. London. 1840.

Bentham, G. Flora Hongkongensis. 18 + li + 482. Reeves, London. 1861.

────── and J. D. Hooker. Genera Plantarum 3:460–636 (Orchideae). Reeve, London. 1883.

Blume, C. L. Bijdragen tot de Flora van Nederlandsch Indië. 1–1169. Batavia. 1825–26.

──────. Tabéllen en Planten voor de Javaansche Orchideën. [10] + plates LXXIII. 1825.

──────. Florae Javae et insularum adjacentium nova series IV. (Orchidaceae). vi + 162 + plates 1–66, 1858. (Contents same as the following entry.)

──────. Collection des Orchidées les plus remarquables de l'archipel Indien et du Japon. 1–162. plates 1–66. Amstelodami. Dates of publication: 1–58, plates 1–22. 1858; 59–114, plates 23–54. 1858–59, uncertain; 115–190, plates 43–66. 1859.

Brown, Robert. Prodromus Florae Novae Hollandiae et Insulae Van Diemen....viii + 145–592 (Orchideae 309–333). London. 1810.

──────. Aiton, Hortus Kewensis ed. 2. 5:188–220 (Gynandria—Monandria). London. 1813.

──────. Lissochilus speciosus. Bot. Reg. 7: pl. 573. 1821 (proposed *Calanthe* R. Br.).

DeWolf, G. P. Jr. On the size of the floras. Taxon 13:149–53. 1964.

Don, D. Prodromus florae Nepalensis. xii + 256. London. 1825.

Dwyer, D. J. Geography and climate. Hong Kong. Report for the Year 1968 pp. 246–55. 1969.

Garay, L. A. and H. R. Sweet. Orchids of Southern Ryukyu Islands. xii + 180. Harvard Bot. Museum. 1974.

Guadichaud, Charles. Histoire naturelle: Botanique. in Freycinet, Voyage autour du monde, . . . Botanique, Orchideae 421–427 (*Luisia, Nervilia* and *Robiquetia*). 1826.

Guillaumin, A. Plantes nouvelles, rares ou criticques des serres du Muséum (Notules sur quellques Orchidées d'Indochine XXIII). Bull. Mus. Hist. Nat. Paris II. 32:186–189 (*Cephalantheropsis* on 188). 1960.

Hance, H. F. Florae Hongkongensis Supplementum. A compendious supplement to Mr Bentham's description of the plants of the Island of Hongkong. Journ. Linn. Soc. Bot. 13:95–144. 1872.

──────. Two new Hongkong orchids. Journ. Bot. Brit. & For. 14:44–45. 1876.

──────. A second Hongkong *Cleisostoma*. ibidem 15:38. 1877.

──────. Orchidaceas quattuor novas sinenses proponit. ibidem 21:231–233. 1883.

──────. Orchidaceas epiphyticas binas novas describit. ibidem 22:364–365. 1884.

Holttum, R. E. Orchids of Malaya (A revised Flora of Malaya Vol. I. 1953). v + 759. ed. 3. 1964.

Hooker, J. D. The Flora of British India (Orchideae). 5:667–864. 1885; 6:1–198. 1894. Reeve, London.

Hooker, W. J. Exotic Flora. 3 vols. I. vii + plates 1–79. 1823; II. plates 85–177. 1825. III. viii + plates 178–232. 1827. (**2**: pl. 138. 1825 *Pholidota*.)

Hsieh, A-ts'ai. An enumeration of the Formosan Orchidaceae. Quart. Journ. Taiwan Mus. **8**:213–282. 1955.

Hu, S. Y. Whence the Chinese generic names of orchids. Am. Orch. Soc. Bull. **34**:518–521. 1965.

————. Dendrobium in Chinese medicine. Econ. Bot. **24**:165–174. 1970.

————. Orchids in the life and culture of the Chinese people. Chung Chi Journ. **10**:1–26. Map 1, figures 1–2. 1971.

————. The Orchidaceae of China. (中華蘭科植物集覽) Quart. Journ. Taiwan Mus. **24**:68–103. 1971; II. ibidem 182–255. 1971; III. ibidem **25**:40–67. 1972; IV. ibidem 199–230. 1972; V. ibidem **26**:131–165. 1973; VI. ibidem 373–406. 1973; VII. ibidem **27**:155–190. 1974; VIII. ibidem 419–467; IX. ibidem **28**:125–182. 1975.

————. Floristic studies in Hong Kong. Chung Chi Journ. **11**:1–25. Maps 1–2, figures 1–3. 1972.

————. Terminology for modern plant taxonomy with Chinese equivalents. 81. Department of Biology, Chung Chi College, Chinese University of Hong Kong. 1972.

————. Sea-shore plants of Hong Kong. Journ. Chin. Univ. Hong Kong. **2**:315–344. 1974.

———— and Gloria Barretto. New species and varieties of Orchidaceae in Hong Kong. Chung Chi Journ. **14**:1–34. 1975.

Humboldt, F. A. von and A. Bonpland. Nova Genera et species plantarum . . . quas in peregrinatione orbis novi collegenut, **1**: xlvi + 302, pl. 1–96. 1815 (281 pl. 85 on *Odontoglossum*).

Jackson, George. Geodorum in Andrews, Botanists Repository **10**: pl. 626. 1811.

Jaeger, E. C. A Source-book of Biological Names and Terms. xxxv + 287. 1954.

Jones, W. Botanical observation of selected Indian plants. As. Res. **4**:231–303 (*Vanda* on p. 293). 1795.

Kränzlin, Fr. Orchidaceae: Monandrae—Dendrobiinae in Engler, Das Pflanzenr. **45(IV. 50. II. B. 21)**:1–382, figures 1–35. 1910; **50(IV. 50. II. B. 21)**:1–182, figures 1–35. 1911. Monandrae—Thelasinae. ibidem **(IV. 50. II. B. 23)**:1–46, figures 1–5. 1911. Monandrae—Orcidiinae—Odontoglosseae Pars. II. ibidem **80(IV. 50)**:1–344, figures 1–29. 1922. Monandrae—Pseudomonopodiales. ibidem **83(IV. 50)**:1–66, figures 1–5. 1923.

Lin, T. P. Native Orchids of Taiwan I. 271. Illustrated. Chiayi, Taiwan. 1975.

Lindley, J. Collectanea botanica. plates 1–41. 1821.

————. The genera and species of orchidaceous plants. xvii + 553. 1830–40. Various dates of publication: 1–94. 1830; 95–134. 1831; 135–158. 1832; 159–225. 1833; 257–335. 1835; 336–366. 1838; 367–379. 1839; 381–533. 1840.

————. Folia orchidacea. An enumeration of the known species of orchids. 1852–59. In 9 parts, dates of publication of genera occurring in China are: *Acampe* 1852; *Calanthe* 1854; *Coelogyne* 1854; *Geodorum* 1854; *Ione* 1853; *Limatodis* 1855; *Luisia* 1853; *Miltonia* 1853; *Odontoglossum* 1852; *Oncidium* 1855; *Sobralia* 1854; *Sunipia* 1853; *Vanda* 1853; *Oberonia* 1859.

Linnaeus, C. Species Plantarum. 2 vols. I. 1–560. 1753; II. 561–1200. (939–954. Gynandria—Diandria.)

Loureiro, J. Flora Cochinchinensis. 2 vols. I. xx + 354; II. 355–744. Lisbon. 1790. (pp. 518–522—Monandria, *Thrixspermum* and *Renanthera*; p. 525—Diandria, Aërides.)

Pfitzer, E. Orchidaceae, in Engler & Prantl, Die natürlichen Pflanzenf. II. **6**:52–224, figures 41–237. 1889.

————. Orchidaceae: Pleonandrae in Engler, Pflanzenr. **12 (IV. 50)**: 1–132. figures 1–44. 1903.

———— and Fr. Fränzlin. Orchidaceae: Monandrae—Coelogyninae in Engler, Das Pflanzenr. **32(IV. 50. II. B. 7)**:1–169, figures 1–54. 1907.

Rafinesque, C. S. Flora telluriana . . . . 4 vols. 1836–38. (**2**:37. 1837. *Pecteilis*.)

Reichenbach, H. G. Icones Florae Germanicae et Helveticae **13–14**:1–194, plates 353–522 (Orchideae). 1851.

————. Orchideae Hongkongenses a cl. Hance et cl. Seemann lectae. Bonplandia **3**: 249–250. 1855.

————. Xenia Orchidaceae—Beiträge sur Kenntniss der Orchideen. 3 vols. I. 1–246, plates 1–100. 1858; II. 1–232, plates 101–200. 1874; III. 1–192, plates 201–300. 1878–1900.

————. Phalaenopsidearum revisio. Hamb. Gart. Blumenz. **16**:115–117 (*Stauropsis*). 1860.

————. Ad Orchidographiam Japonicam Symbolae. Bot. Zeit. **36**:74–76 (*Dactylostalix*). 1878.

————. Otia Botanica Hamburgensia VIII. Novitiae Africanae 79. 1881 (*Manniella*).

Richard, A. Monographie des Orchidées des Iles de Fr nce et de Bourbon. 1–83, plates 1–11. Paris. 1828.

————. *Ludisia* in Dict. Class. Hist. Nat. **7**:437. 1825.

————. Plantes nouvelles d'Abyssinie . . . Ann. Sci. Nat. Paris II. **14**:565–575, plates 15–18 (Orchidées). 1840.

Richard, L. C. Orchides Europaeis Annotationes praesertim ad genera dilucidanda spectantes. Mém. Mus. Hist. Nat. **4**:23–61 (51, *Cephalanthera*; 48, *Platanthera*; 50, *Spiranthes*; 52, *Liparis*). 1818 (reprint p. 39. 1817).

Rolfe, R. A. *Eria fordii*. Gard. Chron. II. **26**:584. 1886.

————. *Pogonia fordii*. ibidem III. **1**:670–671. 1887.

————. *Habenaria rhodocheila*. Orchid Rev. **3**:242. 1895.

————. The *Cypripedium* group. ibidem **4**:327–334, 363–367. 1896.

————. Orchidaceae. Journ. Linn. Soc. Bot. **36**:5–67. 1903. (In An enumeration of all the plants known from China . . . by F. B. Forbes and W. B. Hemsley, also known as Index florae sinensis. 3 vols. 1886–1905.)

Ruiz, H. and J. Pavón. Florae Peruvianae et Chilensis Prodromus. 4 + xxii + 153, plates 37 (p. 120, plate 26 Sobralia). 1794.

Schlecter, R. Die Orchideen, ihre Beschreibung, Kulture and Züchtung: Handbuch für Orchideenliebhaber, Züchter and Botaniker. viii + 836, plates 1–12, figures 1–242. 1914–15. ed. 2. xii + 960, plates 1–16, figures 1–250. 1927.

————. Orchideologieae Sino-Japonicae prodomus. Eine kritische Besprechung der Orchideen Ost-Asiens. Repert. Sp. Nov. Fedde Beih. **4**:1–319. 1919.

————. Das System der Orchidaceen. Notizbl. Bot. Gart. Mus. Berlin-Dahlem **9**:563–91. 1926.

Seidenfaden, G. and T. Smithinand. The orchids of Thailand (A preliminary list). I–IV. **1**:1–98 a–c, pls. 1–2. 1959; **2**:99–184, pls. 3–5. 1959; 185–326, pls. 6–13. 1960; **3**:327–516, pls. 14–21. 1961; **4**:517–647, pls. 22–29. 1963; 648–870, pls. 30–41. 1965. Siam Society, Bangkok.

Schultes, R. E. and A. S. Pease. Generic Names of Orchids. xiii + 331. Academic, N. Y. 1963.

Su, H. J. The native orchids of Taiwan. 138. Illustrated. Harvest Press, Taiwan. 1974.

Summerhayes, V. S. *Microcoelia guyoniana* (Reichb. f.) Summerhayes. Bot. Mus. Leafl. Harvard Univ. **11**:144. 1943.

————. A revision of the genus *Brachycorythis*. Kew Bull. **1955**:221–264. 1955.

Swartz, O. Nova genera & species plantarum; seu Prodromus descriptionum vegetabilium . . . . Indian Occidentalem annis 1783–87 digessit. x + 155 + index (118–126 Gynandria). 1788.

————. Observationes botanicae quibus plantae Indiae occidentalis . . . 1–424, plates 11. 1791.

————. Flora Indiae occidentalis . . . 3 vols. viii + 2018, plates 29 (1391–1564 Gynandria—Monandria). 1797–1806.

————. Dianome Epidendri generi, Linn. Nov. Acta Soc. Sci. Upsal. **6**:61–88 (70, *Dendrobium*; 80, *Cymbidium*). 1799.

————. Afhanding om Orchidernes slägter och deras systematiska indelning. Vet. Akad. Nya Handl. Stockh. **21**:115–138, 202–245 (224, *Neottia*). 1800.

Tang, T. and F. T. Wang. Contributions to the knowledge of eastern Asiatic Orchidaceae I. Bull. Fan Mem. Inst. Biol. Bot. **10**:21–46. 1940. II. Acta Phytotax. Sin. **1**:23–102. 1951.

———— and ————. Orchidaceae, Keys to the subfamilies, tribes, subtribes and genera. Acta Phytotax. Sinica **2(4)**:456–470. 1954. (In Chinese.)

———— and ————. Plantae novae Orchidacearum Hainanensium. Acta Phytotax. Sinica **12**:35–49. 1974.

Thouars (=Petit-Thouars), L.M.A. Histoire particutière des plantes orchidées recuilles sur les trois iles australes d'Afrique. vii + tableau + 32, plates 99. 1822.

Willdenow, C. L. Species plantarum **4**:5–146 Gynandria—Monandria and Diandria. 1805.

Withner, C. L. The Orchids, A scientific survey. ix + 648. Ronald, N. Y. 1959.

————. The Orchids: Scientific studies. xiv + 604. John Willey, N. Y. 1974.

王貴學：《蘭譜》（陶宗儀：《說郛》卷62，頁13-19，清重輯順治三年刊本），ca. 1247.

吳傳澐：《藝蘭要訣》（趙詒琛：《藝海一勺》1933, 卷三，頁1-10），ca. 1811.

鹿亭翁：《蘭易》（《藝海一勺》卷三，頁1-3, 1-5），ca. 1250.

屠用寧：《蘭蕙鏡》（《藝海一勺》卷三，頁1-6），ca. 1811.

趙時庚：《金漳蘭譜》（《說郛》卷105，頁1-10），ca. 1233.

簟溪子：《蘭史》（《藝海一勺》卷三，頁1-7），ca. 1368.

Index

Page numbers in **bold** type refer to the descriptions, and plant names in *italic* type indicate illustrations.

Acampe **110**, 141
  A. multiflora 112
Acanthephippium **90**, 141
  A. sinense 91
Acranthae 6, 127
Acrotonae 11
aerial root 4
Ania **65**, 141
  A. hongkongensis 68
Anoectochilus **43**, 141
  A. yungianus 45
anther 11
Aphyllorchis **54**, 142
  A. montana 55
Appendicula **58**, 142
  A. bifaria 3, 57
Arachnis **103**, 142
  A. flos-aeris 106
Arundina **61**, 142
  A. chinensis 62, 122

Basitonae 11
Bletilla **69**, 142
  B. striata 74
Brachycorythis **29**, 142
  B. galeandra 30, 120
Bulbophyllum **86**, 142
  B. levinei 87
  B. youngsayeanum 122

Calanthe **93**, 142
  C. masuca v. *sinensis* 94
Cattleya **79**, 142
  C. lueddemanniana 81
Cephalantheropsis **54**, 143
  C. gracilis 56

Cheirostylis **43**, 143
  C. chinensis 44, 121
Cirrhopetalum **84**, 143
  C. tseanum 85
Cleisostoma **115**, 143
  C. fordii 118
  C. teres 123
Coelogyne **83**, 143
  C. fimbriata 3, 83
column 10
Cypripedieae 19, 126
Cypripedioideae 19, 126
Cryptostylis **36**, 143
  C. arachnites 39
Cymbidium **96**, 143
  C. ensifolium 3, 99
  C. maclehoseae (cover)

Dendrobium **58**, 143
  D. acinaciforme 59
  D. hercoglossum 122
  D. loddigesii 60, 122
Diandrae 19
Diploprora **104**, 143
  D. championii 3, 108, 123
Disperis **36**, 143
  D. lantauensis 37

endemism 138
Epidendrum **61**, 144
  E. ibaguense 63
Epidendreae 20, 22, 127
epiphytes 2
Eria **88**, 144
  E. corneri 3
  E. rosea 123

  E. sinica 88
Eulophia **69**, 144
  E. sinensis 3, 73

favorable area 137
fleshy root 4
floristic relationship 137

Gastrochilus **114**, 144
  G. holttumianus 114
generic affinity 129
Geodorum **95**, 144
  G. densiflorum 97, 123
Goodyera **46**, 144
  G. procera 48, 121

Habenaria **32**, 144
  H. dentata 32
  H. linguella 120
Hetaeria **49**, 144
  H. nitida 3, 51

Kerosphaereae 20

labellum 8
lip 8
Liparis **72**, 144
  L. chloroxantha 77
  L. macrantha 78
  L. nervosa 122
list of species 133
Ludisia **46**, 145
  L. discolor 47
Lycaste **93**, 145
  L. skinneri 95

Malaxis **79,** 145
   M. acuminata v. *biloba* 80
Manniella **49,** 145
   M. hongkongensis 42
Microcoelia **108,** 145
   M. guyoniana 111
Miltonia **102,** 145
   M. "Belt Field" 102
Mischobulbum **65,** 145
   M. cordifolium 66
Monandrae 19
monopodial stem 5
morphological diversity 129

Neottieae 19, 20, 126
Nephelaphyllum **61,** 145
   N. cristatum 64
Nervilia **36,** 145
   N. fordii 38

Odontoglossum **98,** 145
   O. grande 101
Oncidium **98,** 146
   O. varicosum 100
Ophrydoideae 19
Orchideae 19, 20, 126
Orchidoideae 19, 126
Ornithochilus **116,** 146
   O. eublepharon 119

Pachystoma **69,** 146
   P. chinense 70

Paphiopedilum **26,** 146
   P. purpuratum 27, 120
Pecteilis **27,** 146
   P. susannae 28, 120
Peristylus **34,** 146
   P. goodyeroides 121
   P. spiranthes 34
Phaius **90,** 146
   P. tankervilliae 92
Phalaenopsis **104,** 146
   P. amabilis 109
Pholidota **82,** 146
   P. cantonensis 3
   P. chinensis 82
phytogeographic significance 136
Platanthera **29,** 147
   P. mandarinorum 3
   P. minor 31
Pleuranthae 6, 127
pollinia 11
Polychondreae 19
population paucity 131

Renanthera **103,** 147
   R. coccinea 107
Resupination 7
rhizome 5
Robiquetia **115,** 147
   R. succisa 117
rostellum 11

Sophronitis **89,** 147
   S. grandiflora 89
Spathoglottis **71,** 147
   S. fortunei 75
   S. pubescens 75
Spiranthes **40,** 147
   S. hongkongensis 41
   S. sinensis 41
symbiosis 4
sympodial stem 5
synsepalum 8

Tainia **65,** 147
   T. dunnii 67
terrestrial orchids 1
Thelasis **71,** 147
   T. hongkongensis 76
Thrixspermum **103,** 147
   T. centipeda 105
Tropidia **53,** 148
   T. hongkongensis 53

Vanda **110,** 148
   V. teres 113
velamen 4
Vrydagzynea **52,** 148
   L. nuda 52

Zeuxine **46,** 148
   Z. gracilis 50, 121

MAJOR LIFE EVENTS OF PROFESSOR SHIU-YING HU

Text and photos courtesy of Shiu-Ying Hu Herbarium

1908–1925

"I was born in the Qing dynasty, and have experienced three different eras."

Professor Hu with her mother. Her hometown is located in Yuanjiawa in Xuzhou, Jiangsu province. Her parents named her Xinglian (興廉).

Professor Hu (second from the right) received a scholarship to the Genuine Heart School in Xuzhou, a boarding school run by missionaries.

1926–1933

"A small mustard seed, sown in the fertile farmland of Ginling."

Prof. Hu received a scholarship from the Genuine Heart School to study at Ginling College in Nanjing (1926–1933).

An aerial view of the campus. When Prof. Hu was in her second semester at Ginling, an introductory course on biology was offered by three newly recruited professors in the college. The physical observation and fieldwork she learned there broadened her vision, and also aroused her curiosity about nature. The singing birds, blossoming flowers, croaking frogs and chirping insects were all encouraging her to observe and explore. As a result, she changed her mind and instead of majoring in physics, she decided to become a student of biology.

The introductory course, called "How to Study," given by the Vice President of Academic Affairs, Ms. Vautrin, guided Prof. Hu in her studies.

Dr. Cora Reeves (standing on the right) was a brilliant teacher in the Department of Biology. She was not good at speaking, but was excellent at organizing. She led the students outdoors and inspired their ability to observe nature with their eyes, ears, and minds, and to organize their thoughts.

Left: Prof. Hu's beloved mentor, President of Ginling College, Ms. Wu Yi-fang.
Right: Chairman of the Department of Sports, Ms. Cui Ya-lan.

Being a biology major at Ginling College, Prof. Hu always remembered each plant and animal there. And as an athlete active in different sports at that time, she also fondly recalled the sports field.

Prof. Hu (second row, first from the left) worked as a sports teacher in the Genuine Heart School for two years during her studies at Ginling College.

"**When I arrived in Guangzhou, my knowledge of tropical plants was very limited. The demands of teaching motivated my progress in learning. And within two years, I became an expert on plants in South China (the Lingnan region).**"

1934–1937

學大南嶺
LINGNAN UNIVERSITY

Guangzhou is where Prof. Hu received her foundational training in botany.

The campus of Lingnan University.

Graduating with a Master's degree in science, Lingnan University. Prof. Hu is the second from the left.

1938–1946

The doors and windows of the medical building of West China Union University were destroyed by a Japanese air raid on June 11, 1939.

Prof. Hu worked as a lecturer of botany in the Department of Biology at West China Union University from 1938 to 1946, teaching plant taxonomy and general botany.

Students of Ginling College performing a folk dance. During the War of Resistance against Japan, Prof. Hu organized charity sales and performances and raised 17,000 yuan for two new projects—buying yarn to knit sweaters for soldiers and providing additional free beds for hospitals.

1946–1968

Front gate, the Department of Biology, Harvard University. Prof. Hu entered these gates for the first time on September 20, 1946.

Prof. Hu got a full scholarship to Radcliffe College, a women's liberal arts college that was the coordinate institution for Harvard College, which was all-male at that time.

Prof. Hu's mentor, Prof. E. D. Merril, she had longed to study with this famous botanist back when she was studying at the Lingnan University. Merril was director of the Arnold Arboretum from 1935 to 1946 and Arnold Professor thereafter until his death in 1956.

Staff of Prof. Hu's Flora of China Project, taken before the University Herbarium in 1956. Professor Hu took the photo. Professor Hu stayed on at Harvard for more than 20 years after obtaining her PhD. These years were spent in intensive research at the Arnold Arboretum.

The Chinese Christian Church of New England, in Boston's Chinatown. Prof. Hu was an elder of the church.

In 1960, Prof. Hu bought a house in suburban Boston, which she named *anding tang*, meaning "a place to settle down." She provided accommodation for many people from Ginling College who were visiting Boston.

1968–2012

"I had a wish in my mind before coming to Hong Kong that I hoped to experience more Chinese culture here."

Dr. and Mrs. Andrew Roy were lifelong friends of Prof. Hu. Dr. Roy later became the dean of the Divinity School of Chung Chi College, CUHK in 1964. He invited Prof. Hu to teach at the Department of Biology, CUHK in 1968.

Chung Chi College follows in the tradition of the 13 Christian universities in mainland China, such as Ginling College, Lingnan University and West China Union University, which were significant in Prof. Hu's life.

Prof. Hu collecting specimens of the Pitcher Plant in Lung Kwu Tan, Hong Kong, 1968. Prof. Hu considered that the most efficient period of teaching and learning in her life was during her stay in Hong Kong.

Prof. Hu explaining plant specimens to the public on the Open Day of The Chinese University of Hong Kong's anniversary, 1969.

On a joint field trip with her Ecology and Plant Taxonomy courses in 1969.

Prof. Hu and some staff from the Department of Biology, 1970.

Mrs. Gloria Barretto, a Portuguese woman, was fond of studying wild orchids in Hong Kong. Every day after work, she would take a blossoming orchid and a cup of milk tea to Prof. Hu's office. After teatime, Prof. Hu would teach her about orchids by observation and dissection (using needles and forceps), and morphological explanations. After a few years, they had documented more than 100 species of native orchids, with detailed information on their leaves, roots, and flowers. More than 10 new species were published, all with authentic scientific illustrations.

In the Herbarium at the Department of Biology, The Chinese University of Hong Kong, October 1993.

Shiuying Bamboo, an endemic species in Hong Kong, which can be found on the way to the Hua Lien Tang student hostel at CUHK.

"When I think back on all the mountains I have climbed in the world, there are only two that I climbed up from the east, west, north and south. One is Purple Mountain in Nanjing, and the other is Ma On Shan in Hong Kong."

On February 2, 2002, Prof. Hu led students on a campus walk to learn about the plants at Chung Chi College.

In August 2005, Chung Chi College renovated a ground-floor apartment at Lee Shu Pui Hall (a student hostel) for Prof. Hu, who had difficulties with mobility. A nice living room and an access ramp for a wheelchair were specially built for her, so that she could enjoy her later years in comfort.

On February 10, 2006, to celebrate Prof. Hu's 100th birthday, Healthworks established the Hu Shiu Ying Award for Chinese Medicine Students during Chung Chi College's Assembly.

In 2009, an American sculptor, Mr. Matthew Buckner, carved a statue of Prof. Hu, which is now kept at the Shiu-Ying Hu Herbarium.

In May 22, 2012, Prof. Hu left us at the age of 104.

In 2013, at the opening ceremony of the Shiu-Ying Hu Herbarium.

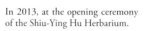

"Chung Chi is my home in China."